Money in Rhode Island Reds

by De Graff Poultry Farm

with an introduction by Jackson Chambers

This work contains material that was originally published in 1913.

This publication is within the Public Domain.

This edition is reprinted for educational purposes and in accordance with all applicable Federal Laws.

Introduction Copyright 2017 by Jackson Chambers

Self Reliance Books

Get more historic titles on animal and stock breeding, gardening and old fashioned skills by visiting us at:

http://selfreliancebooks.blogspot.com/

Introduction

I am pleased to present yet another title in the "Chicken Breeds" series.

The work is in the Public Domain and is re-printed here in accordance with Federal Laws.

Though this work is a century old it contains much information on poultry that is still pertinent today.

As with all reprinted books of this age that are intended to perfectly reproduce the original edition, considerable pains and effort had to be undertaken to correct fading and sometimes outright damage to existing proofs of this title. At times, this task is quite monumental, requiring an almost total "rebuilding" of some pages from digital proofs of multiple copies. Despite this, imperfections still sometimes exist in the final proof and may detract from the visual appearance of the text.

I hope you enjoy reading this book as much as I enjoyed making it available to readers again.

Jackson Chambers

DeGRAFF'S BOOK ON REDS.

This book, my traveling salesman for 1913, and incidentally the finest poultry catalogue ever published on earth, I dedicate to my dear departed Mother, to whom all credit is due, as whatever originality I may possess is purely heriditary. All original ideas herein advanced are entirely of my own developing and I hope there may be something that will help my fellow man.

E. T. DeGRAFF,
Author.

My Father   My Son   Myself

As I am collecting material to write a family history of the DeGraff family in this country, I would like to hear from all DeGraffs or descendants from DeGraffs, that I may make it as complete as possible. I have my son's pedigree complete for nine generations in this country, and for ten generations on the other side, which may be of interest to descendants of this family. We have owned this farm from the original grant and I hope to develop the village of "DeGraff" on this farm which will perpetuate the name forever.

Leading Red Specialist of America. S. C. and R. C. Rhode Island Reds

THE CHAMPION AMSTERDAM WINTER LAYING STRAIN

De Graff's Book on Reds is International Authority on Reds

MY MOTTO: "YOUR MONEY'S WORTH OR YOUR MONEY BACK"

Bell Phone No. 94-J Cable Code--Redgraff

Amsterdam, N. Y., March 15, 1913

Brother Fancier:

In preparing my 1913 catalogues, I have tried to give you as true an idea of this farm, the quality of Red stock raised here and my methods of caring for them, as the limited space will allow. I will prove to your perfect satisfaction that the matter of distance makes no difference in shipping eggs or stock, when properly handled, as I show by testimonials that I have shipped to over half the Earth with perfectly satisfactory results, so that I feel safe in saying I could guarantee delivery to any civilized country of the world.

My poultry business has grown to such magnitude that in 1908 I resigned from the Farmers National Bank, which I had been connected with for 20 years, that I might give this business my undivided attention. My plant represents the results of years of hard work, unlimited expense, and careful study, in building up this strain of fowls, which I have through live advertising, made known all over the World, as "DeGraff's Vitality Reds," "The Best Reds in America" "The Greatest Utility Fowls on Earth."

I have moved my whole plant to a large apple tree orchard of 600 full grown trees, which gives me unlimited range for all my best breeding pens, thereby giving them conditions, true to nature, and assuring perfect health in fowls and fertility in eggs which can not be duplicated under any other conditions. After 25 years of practical experience with over 20 varieties of most popular fowls, I have decided the Rhode Island Red fowls are "The Greatest Money Makers" and hereafter I will breed only Single Comb and Rose Comb Reds in as large numbers as possible to properly handle, and defend my position of "The Leading Red Specialist of America" by keeping my stock as complete as possible, at all times of year, that I may be able to sell you Reds of any age, of either variety, and give you extra good value for any priced bird you care to buy.

Make all remittances payable to

Ed. T. DeGraff.

(This Catalogue is Copyrighted 1913 by Edw. T. DeGraff, Amsterdam, N.Y.)

"FOR WHOM THE LORD LOVETH HE CHASTENETH AND SCOURGETH EVERY SON HE RECEIVETH"— HEBREWS XII-6

This text, full of truth and consolation, has been a great help to me for several years back, as only those driven by necessity accomplish real work in this world. Blessed and useful is the lash that drives us to better deeds. Sometimes the lash strikes from within, and sometimes from without. The study of human nature, in everyday walks of life, is more interesting to me than most vaudeville performances on the stage. There is a natural tendency in the human race, like sheep to follow the leader, and often I fear there is about as much thought behind it. The number of real men that will stand for what they know to be right against public opinion is very small, and should be appreciated. While I consider nearly every poultry editor in the country is my friend, and all fair-minded ones have admitted they never received any complaints from any of their subscribers but what I made good promptly, still, when Grant M. Curtis called them together, after I was pushed over Niagara Falls, and maliciously and falsely stated that he had received a wheelbarrow full of complaints against me, there was no one that had independence enough to call his bluff and deny what they must have known to be false.

Now that I have heard all the inside particulars about this poultry magnate meeting, and all the poultry papers that I care to advertise with have asked for my advertisement, I will state for the information of those interested that I stopped my advertising with Curtis' paper before the A. P. A. meet was ever held, as it was not paying, so he never had the opportunity to refuse my advertisement; hence his apparent reform loses much of its true ring, and personal grudge appears very strong to any fairminded person, and my daily mail proves the country is full of this class.

I offer the following editorial comments taken from the leading papers of the country to prove my contention for the last four years, had more truth in it than any report published heretofore. Taking the testimonials on page two of supplement and these editorials, I don't see how I could possibly give any stronger evidence of my reliability and ability to fill your orders satisfactorily.

"Apparently a campaign is on to reinstate E. T. DeGraff in the American Poultry Association. Several of the exchanges received so far this month contain DeGraff's advertisement, good reading notices for DeGraff, and editorials advocating the lifting of the ban. I never approved the extent of the penalty visited upon Mr. DeGraff. He provoked the extreme severity of the sentence, not by his faults, but by his attitude toward the Executive Board considering his case. It always seemed to me that the judges, presumed to be disinterested, considering all the circumstances should have overlooked this.

"But I question whether the method of working for a reinstatement is a good one. It does not look to me like good policy doing business. I'd much rather live outside as a shining example of the ineffectiveness of its ban than be restored to membership and unanimously elected President of the organization."—John H. Robinson, Editor Farm Poultry, Boston, Mass.

It is beginning to look to us, and we know we are not alone in the belief, that Mr. DeGraff is being punished for having enough good American spirit to defend himself.

To keep the finger of suspicion pointed at a man for four or five years is a punishment which should be handed out only in extreme cases. By no stretch of the imagination can this be justified in this case. Mr. DeGraff is in no sense of the word a suspicious character, and the cloud of suspicion which the A. P. A. is holding over his head should be lifted at once.

Poultry carried DeGraff's advertisement many years before this trouble came up and during all this time we never had a complaint from one of our readers who bought stock or eggs from him. His customers found that he was giving them good value, and people began to wonder if the "Wise Guys" would try to keep him in outer darkness forever. Ed. DeGraff is "coming back," and now and then you will find an editor who dares say a word in his behalf, and are asking for his advertisements again.—Editor Poultry, Poultry Publishing Co., Peotone, Ill.

Among the Rhode Island Red advertisers appearing in the February number of the Inland is Edward T. DeGraff, of Amsterdam, N. Y. For years Mr. DeGraff has been working under a handicap, and a serious handicap. In addition to this trouble with the American Poultry Association, his wife was thrown from a buggy and was injured. That would have taken the heart out of any man, as he has been compelled to have a trained nurse with her for a number of years. And then to add to his other troubles, on his return from one of the shows eight of his barns and buildings were ablaze and were completely destroyed. But through it all Ed. DeGraff came up smiling, and we hope to see him succeed in his chosen profession as a poultryman, and with the number of good birds we have found in different parts of the country from his yards, we know that he breeds a lot of good ones.—Theo Hewes, Editor Inland Poultry Journal, Indianapolis, Ind.

I for one could never see the justice of a life sentence for the crime committed.

I furthermore believe that Mr. DeGraff in the past ten years has been largely instrumental in making the Rhode Island Reds one of the most popular of all breeds, and notwithstanding the fact that he was expelled from the A. P. A., even this didn't cool his ardor in pushing the Reds to the front, and this was done in the face of adverse circumstances.

After three years of careful thought and consideration, it seems to be the general impression that the sentence was too severe for the crime which Mr. DeGraff was convicted of.

Mr. DeGraff has done much, not only for the Rhode Island Reds, but for the advancement of the Poultry industry in this country.

I have known Mr. DeGraff almost from the time that I have been connected with a Poultry Journal, and must confess that in handling his business in the past I had very little complaint concerning his business relations with his customers, nothing more than normal, as usually follows in the wake of a large business.

I have visited Mr. DeGraff at his home and have looked over his flock of birds and cannot help but say at this time, as I did in the past, that I was greatly surprised at the extent of Mr. DeGraff's plant, seeing so many birds of their quality.—F. W. DeLancey, Editor Poultry Fancier, Sellersville, Pa.

We believe that everybody who is at all familiar with the Rhode Island Red history for the past ten years will admit that it was largely through Mr. DeGraff's efforts that this variety came to be the popular fowl that it is today, and in our opinion this is further reason why his advertisement appears in this issue.

There is no doubt but what DeGraff has made some serious errors and done that which would have been better left undone, but who is there among us who has not made errors?

Therefore let us be charitable and trust Mr. DeGraff in the future will profit by the errors made by both himself and others in the past.—James W. Bell, Editor American Poultry Journal.

There is no doubt but what Mr. DeGraff committed an act of wrongdoing (exhibited one faked bird), in fact, he admitted exchanging birds, and therefore should have been duly, but justly, punished. The offense, however, was no greater than that of others implicated with him that were let off without reproof. Furthermore, there are dozens of other breeders and exhibitors, all A. P. A. members, that have transgressed from the good A. P. A. laws quite as badly as ever DeGraff did, but not a hand has ever been raised against them. Whether or not DeGraff, under the circumstances, was deserving of the penalty visited upon him, he has at least been sufficiently punished. He is honest at heart, a skillful breeder, and an enthusiastic, aggressive worker, the kind of man the poultry industry needs quite as bad as he needs the support of the A. P. A. He has specialized in Reds for many years, and by his novel methods of advertising, handsome catalogues and beautiful color supplements, which have been widely distributed, he has done as much, if not more, than any other man to bring about their present day popularity.—D. M. Green, Editor Poultry Husbandry, Waterville, N. Y.

I am willing to stand for the old sayings, "You can always trust a man that likes flowers," or "You can judge a person's private character by his dooryard." This is my residence and west side yard.

My son John is the eighth generation of DeGraff's to live on this farm, and we have the original deed practically from the "Original Reds" hanging in the hall, so that we are beginning to feel very much at home here.

HOME SWEET HOME.

Never in the history of the world have we heard so much about the high "cost of living" together with the cause and cure for same and heard so much said in favor of this "Back to the Soil" movement as we have this last winter.

Any student of political economy, cannot help but see that the cities are over crowded with unemployed, who are merely existing from day to day, while the country is full of good farms that are simply going to waste for lack of the proper men to take hold and make them a paying investment and incidentally make a home worth living in, and a fit place to bring up a family of children, that will be a credit to their country, instead of being raised in a hot bed of vice as many portions of our large cities are getting to be.

There is no business that offers so many opportunities for making a comfortable living as the many branches of the poultry industry, and no great amount of land or money is required to get a start, that will gradually develop by judicious management and thrift into a business that knows no boundary, as the supply of strictly fresh eggs and prime broilers has never been over supplied, and as years go by and the value of poultry and eggs as nourishing food, is better understood, the demand will be double what it is today.

Think of the number of families that have gone without eggs all this winter when they could easily have kept a few hens in their back yards and not only supplied the table with eggs, from the kitchen scraps, but furnished employment for the children's spare moments and kept them out of mischief, and incidentally give them some idea of business by having them manage the poultry department for you.

I have been through all stages of city life, and I can unconditionally recommend all home seekers to select a suburban home with as many city conveniences as possible, but above all get enough land that you can have your own fowls, and possibly a horse and cow, that you may enjoy the little daily pleasures of life that make it worth living. It is not the great big events in life that count most, but each little daily pleasure, that goes to produce some grand result in the end, and there is something wrong with the person that does not enjoy seeing the wonderful workings of Nature, as each season unfolds its beauties.

Where is there any greater satisfaction in life or pardonable pleasure than to see the daily development of your children, as they mature in perfect health to that type of citizen this country is so sadly in need of.

PRACTICAL POULTRY CULTURE TAUGHT FREE

I will furnish board and lodging to anyone wanting to learn the poultry business, as it is managed under most advanced methods, and give them the benefit of my experience, free of charge. I can give references of several graduates that are now filling important positions in this country and Europe, while a recommend from this farm and a certificate from Cornell has two years landed the best job of the year from this institution.

As I will be building an entirely new plant this spring on new lines anyone can get an experience not to be gained anywhere else, and will be invaluable to anyone intending to build for themselves later.

HOME COMFORTS OF COBBLESTONE FIREPLACE

There is a growing tendency to include the old-fashioned fireplace in the best bungalow homes of today.

I designed and built this fireplace on entirely original ideas, whereby five rooms are heated and the whole house ventilated, and a most artistic effect secured. Anyone in search of an ideal country home on above lines should write me, as I am in real estate business and can sell you anything from a 50-foot building lot to the above residence.

My "Block of Ages," Foundation of My Success in Breeding and Advertising.

To anyone that has ever raised Red chickens alongside of others I need not explain that there is no chick hatched that will compare with the Red for vitality right from the start, and if given a decent chance to live at all they are sure to grow and surprise you by their rapid development of bone and flesh.

You can see the vitality in every specimen shown and the plump breast and well developed thighs make them a delicious morsel even at six weeks. Chicks hatched from my free range stock will live and grow like these, and not die like the chicks from confined, hot house, stock, we read much about now dying from "White Diarrhoea."

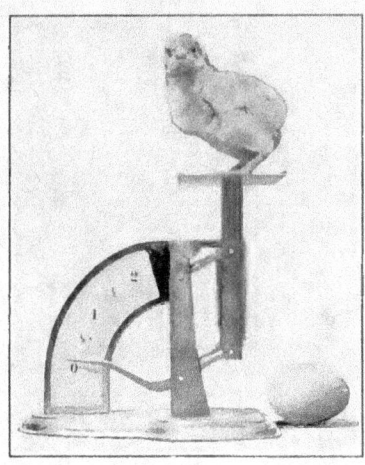

A chick is not entirely incubated until about 3 days old, during which time he should be kept warm and quiet and not fed or watered. He will lose but slightly in weight and get a perfectly healthy start, and a chick once started right is half raised.

MY 1909 PUBLICITY CAMPAIGN ARTICLE.

This article on vitality of Reds was published in every paper of any consequence in the U. S. during summer of 1909, and not only introduced the many good qualities of this breed to thousands of amateurs, but was a benefit to every breeder of the breed.

The grouping of chicks in this picture is an original idea of my own which has never been seen before, and illustrates plainer than words what I want to explain.

VITALITY OF YOUNG RHODE ISLAND REDS.

Without doubt the greatest problem before the poultry fraternity today is the question, why so many chicks die during the first few days of their existence, and how it can be avoided. One of the greatest pleasures to be derived from this business is "anticipation," and the worst "realization" is a bad case of what is commonly called "White Diarrhoea," which in many cases is no diarrhoea at all, but a weakness in the lungs and other internal organs. The ability to perpetuate your flock each year with larger, stronger, more prolific specimens is the cornerstone of success in this industry, and the breeder that is a successful chicken raiser can make more profit from poultry, either from the utility or fancy standpoint, than any crop raised on the farm, especially when conducted under favorable circumstances.

The poultry industry is this nation's "King Business," and when we realize that the average life of a fowl is less than one year we see the importance of successfully rearing the young each year. My practical experience dates back to when I had 100 setting hens going all at one time as a side line, while I was going to school, and later when I took up incubators and brooders I found I really knew less each batch I took off, and when my hatching was done by experts I was worse off than ever.

When I resigned from the bank, after being tied up with the filthy lucre for twenty years, I took active management of the business and made the raising of young chicks my hobby and carried on many interesting experiments which proved very beneficial to me, if not the chicks. Owing to the fact that I had trouble made me give the subject all the more careful thought at every stage, until I felt I was on the right track.

There can be no doubt but that washing eggs laid by unhealthy fowls can be improved by washing in alcohol or by using a solution of 1-2000 of bichloride of mercury or corrosive sublimate, as this will kill all germs that might be deposited on the eggs coming through the same organ that all other secretions pass through. It can not injure the eggs and it has been known to produce very satisfactory results.

You will never see a hen steal her nest the second time in the same place, so don't ever set your incubator without first thoroughly disinfecting with zenoleum, creoside or some other good tar disinfectant.

I found, while all the latest rules for disinfecting, sterilizing, and control of heat, were more or less beneficial, under some circumstances that there was that indescribable something that nature used in connection with the hen that no artificial method could duplicate, and that unexplainable quality called vitality played a very important part in results obtained from any system. I found that some chicks, if properly incubated at all, would live almost regardless of how brooded or fed, while other batches which might appear perfectly normal up to three days old were just as sure to die inside of two weeks as the sun is sure to rise. No matter how brooded or fed whether under hens or not, they are bound to die, as they are imperfectly developed specimens and their organs will never do the work that is expected of them, and even though they live a few months they will always be a poor investment to anyone in the end. The more I know of the poultry industry the more I realize its wonderful possibilities and the unlimited opportunities yet to be discovered. Where is there a more interesting problem in nature to study than the development of life in the egg by artificial or natural methods? This small, undiscernable speck we call the germ of a properly fertilized egg hangs by delicate ligaments near the yolk of the egg waiting for the proper degree of heat to start its development, and when given a certain number of heat units it passes through all the interesting stages and we have the perfect chick. The process can be seen through the shell to a certain extent and the various degrees of vitality are remarkably demonstrated during this time. The strongest germs show strong from the first sign of life and their development shows strength at all stages, until it is time to come forth, when the expert can tell by handling which eggs will hatch the perfect specimens that will live. The weaker germs gradually die off at all stages of incubation, according to the vitality they possess, even up to the hatching time, and even a few days beyond, as the chick is not fully incubated until the internal nourishment called the unabsorbed yolk is taken into the system, and he is completed ready to be nourished in the regular way, and the chicks that have to be helped out of the shell are seldom worth the time taken to rescue them, unless hindered by some unnatural cause.

Think of the wonderful examples of vitality of the germs which have been demonstrated by long shipments of eggs for hatching to nearly all parts of the world with comparatively good results. I recently shipped 30 Rhode Island Red eggs to Alaska on an order, hardly expecting him to get much if any returns under the circumstances when what was my surprise to hear that he had hatched 24 chicks and raised 23 of them, and he was most strongly impressed with their vitality up there. I have found that the express companies are directly blamable for many poor hatches received from shipped eggs on account of their reckless handling, which jars the eggs so severely that the little ligaments that support the germ in place are broken and the future development of the germ ruined although the egg might have been perfectly fertilized and the shell not broken in the least.

Any method of relieving the jar of the cars and handling will greatly increase the chances of hatching and I find a pneumatic cushion under the package helps wonderfully on long shipments. There is no possible doubt but that hen hatched chicks are superior to incubator hatched chicks, although there is a great difference how each incubator is managed and under what conditions they are run. While the vitality has much to do with it, still the strongest germs can be utterly ruined by improper incubation, and it will come forth, if at all, with tuberculosis in one form or another, only to die, and probably the brooder or system of feeding will be blamed for what no power on earth could avoid.

Actual photograph of red chicks 5 minutes old.

Think of this little ball of feathers cramped up in its small space forcing its way through the hard shell by its own efforts, from the inside, and how each struggle accomplishes some slight step towards liberty until it gives one strong effort and the shell opens and it comes forth entirely of its own efforts, and after a few minutes' rest, during which time it dries off from the heat of the hen, you see a fluffy, bright-eyed, strong legged chick which appears to be twice the size of the egg from which it has just escaped.

Take the normal hen hatched chicks, does it hurt them to crawl out from under the hen and get chilled, and even fall out of the box, and even when found apparently dead the hen's warmth will bring him back to life as good as ever, if the vitality and perfect development is there behind it? No healthy chick needs to be taught anything as they know intuitively how to eat, drink, scratch and find their way back home right from the start. While in one sense they are about as capable of caring for themselves as any young creature I know of, still in other respects I know of no one having so many natural enemies, ranging from the little louse to the human louse (who sometimes reaps the reward of a long season's watchfulness) and incidentally the rats, cats, dogs, skunks, minks, hawks, crows and foxes which are always ready to improve any opportunities you give them.

While tonics or antiseptics may help guard off indigestion and create appetite in young chicks, still I believe the best remedy is plenty of green food right from the start, and give them all they can eat, either by free range or supplying from local beds.

You will find as your chicks get about six weeks old sometimes that they seem to lose their appetite and get weak, which you can help by putting a few quassi chips in their drinking water at times, and if you once used my drinking bottle with iron base you would find it so far superior to all others that you would use no other. The iron acts as a tonic.

Fowls in their native state eat large quantities of green feed of all kinds along with all kinds of insect life, which accounts for much of the strong vitality of eggs laid by such fowls, and the difference between two flocks of chicks, one raised with green feed and the other without, is very marked, as the batch with green feed will walk right away from the others, and the more or less stunted condition of the other flock, together with the white feathers which they will have from lack of proper material to make the necessary coloring matter to carry out what nature intended them to be from their breeding are simply the result of unnatural conditions.

You can't take something from nothing, and even nature will not stand for this method, and three generations of fowls bred under unfavorable conditions and hatched by unfavorable methods will not perpetuate themselves, showing the system is decidedly wrong from the foundation. Green feed can be furnished in many ways by a little management by raising lettuce, turnip tops, sprouted oats and other grains, cabbage and common beets, but the best of all is the mangel-wurzel beet, which can be kept all the year round, and very easily fed at all times to all kinds of stock. Naturally there is a great difference in different breeds of chicks about their vitality and characteristics, and I can honestly say that the Rhode Island Red chicks are the hardiest, strongest, hustling chicks of them all, and they will live if given a chance to hustle for themselves.

Their origin makes the vitality still show to a marked degree and the accompanying pictures tell the story far plainer than any words can describe and shows their many good practical qualities at all ages.

The 1¼ lb. Saratoga broiler is plump and tender, while the 2 lb. broiler at two months can't be compared with by any other breed, and the 3 lb. and 4 lb. roasters have those plump yellow bodies and yellow legs that command the highest prices.

Just what it is and how to secure this desirable vitality is something that would be hard to explain definitely, but I find that the more we follow nature and the less we try to improve on nature's ways the stronger vitality we have, and the more we hurry things unnaturally the weaker we are making our stock. Give a pen of Reds plenty of pure food, fresh water and fresh air houses, protected from exposure and drafts, and you will not be fighting at every turn the natural results of unnatural methods of poultry culture.

> There is so much that is bad in the best of us,
> And so much good in the worst of us,
> That it doesn't behoove any of us,
> To talk about the rest of us.

SUMMER HOME OF DeGRAFF'S VITALITY REDS. 600 APPLE TREE ORCHARD, COLONIZED WITH MOVABLE HOUSES, WITH WATER PIPED TO ALL PARTS. A PERFECT POULTRY HEAVEN ON EARTH

THE HAPPY HOME OF THE BEST REDS IN AMERICA.

This farm of 75 acres of fertile Mohawk Valley land, capable of producing everything that is needed to make a successful poultry farm, is situated one mile east of Amsterdam, N. Y., on the "Main highway around the World," so we are situated to advantage for shipping facilities to all parts of the country.

The poultry buildings are situated on the North side of the N. Y. C. & H. R. R. R. and in sight of the West Shore R. R., and one of the best equipped interurban electric roads in the country has a private stop right at house, so that any visitor can take accommodation cars from either Amsterdam or Schenectady and step off right at the yards and need not lose over one hour if on a hurried trip to Boston or New York City. Nothing gives me more pleasure than to have a call from one of these skeptical fanciers that has been pumped full of misrepresentations about my stock and myself, by my esteemed contemporaries, as I really enjoy the opportunity of proving by facts, that all I claim for my stock in my catalogue I can prove in cold facts here. After many years of experimenting in a practical way with all different methods of housing and caring for poultry I now have the opportunity to lay out a poultry ranch on entirely new ground with entirely new houses designed as modern as money and brains can make them, and all so naturally favorably located that visitors say it seems more like a dream than stern reality, and when I have all my plans completed, I honestly believe I will have the best equipped plant for breeding of Standard bred poultry to the highest state of perfection that there is in existence.

I will eliminate all that farce and misrepresentation of facts that constitutes such an important part in the poultry industry, as controlled by the A. P. A and establish a practical poultry plant, that will be known all over the world as the Leading Red Specialist of America, and the Happy Home of the Greatest Money making fowls on Earth.

For the benefit of the 600 full grown apple trees, I had the field which was formerly seeded down to clover, plowed four furrows in each direction between all trees, thereby opening up the ground for better nourishment to the feeding roots of the tree all the way round. I sowed each strip running east and west with rye which is now coming out fine for early green feed, and as soon as the ground is fit I will sow the strips running north and south to Dwarf Essex rape, at different periods during the summer and harrow it in, thereby affording more green feed of three very important kinds than they can possibly eat, so there will be enough fodder left to keep some cows to furnish milk for the young growing stock, and there is nothing that helps put the finishing touches on young growing prize winners than a liberal feed of milk in many forms.

Nature could not have done better for me in the water supply, as I only had to go about 300 feet north of the orchard, where I found a vein of water that had not been dry for years, and by building a cement tub for same, housed over, I have a liberal supply of fine spring water that will supply water all the year round.

We have the water piped to the center of orchard and soon have it running all the way around the driveway with faucets or drips for each house or combination of houses.

I will continue to build all summer, making ideal conditions for thousands of Reds, and as they are all of one breed and each cluster of four houses sufficiently far enough from the next bunch that there will be no need of fences, and all fowls will forage at their own free will among the leaves, under trees, and incidentally fertilize the trees, and kill all insects that might damage the fruit and trees.

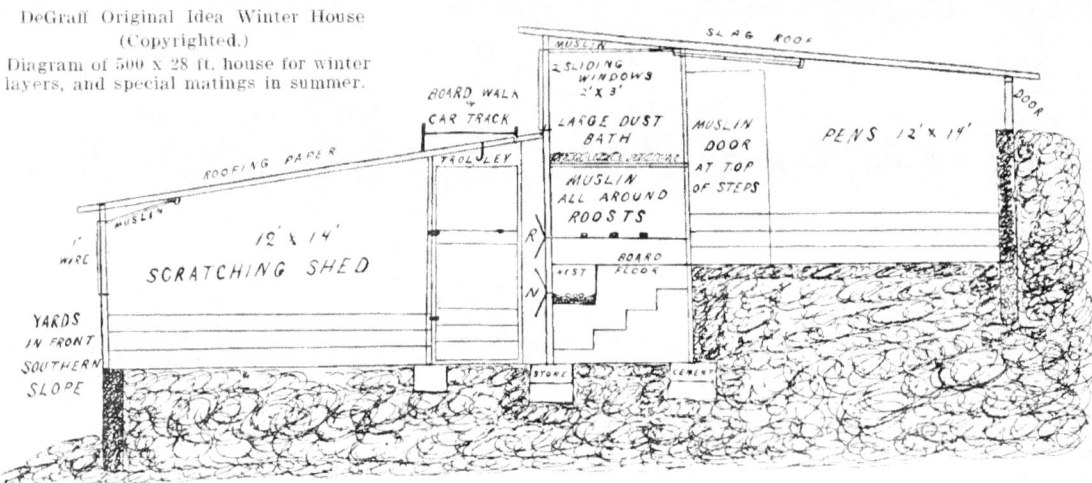

DeGraff Original Idea Winter House (Copyrighted.)
Diagram of 500 x 28 ft. house for winter layers, and special matings in summer.

ORIGINAL IDEAS IN POULTRY HOUSES TO BE FOUND ON THIS FARM

This diagram gives the ideas that I have adopted in my large long house at the south side of the orchard with yards running down the southern slope, all planted with pear and plum trees and seeded with alfalfa before the fowls are put in. I hope you will carefully study the many new original practical ideas that are shown here for the first time, and many are very important labor saving points, that mean money saved every day.

This house is intended for the majority of the fowls when in winter quarters by having about 50 layers in each double section, and in spring it can be easily divided into front and back sections for smaller breeding pens.

The overhead boardwalk and car track has many uses and costs but very little, as I intended to use it for feeding both front and back pens with scratching feed, and heavy loads of sand can be delivered right to the dusting baths and straw and other material brought right where it is needed, and the refuse delivered to the proper shed for this purpose.

The overhead trolley in doorway passage can be utilized the same as if it was a separate aisle, by keeping all doors to the rear part closed in morning until the work is done and throwing all front doors open automatically from the feed room.

In fact, every window and door will be worked by wires and levers from the feed room and all pens watered summer and winter from the feed room without going to any part of the house, with a system shown in my catalogue.

All eggs can be hunted from this aisle, and the manure removed right into the trolley car, in one operation at each pen, and this same car can bring sand and straw, and remove the refuse without any back-breaking operations and in quarter of the time.

The large dusting bath, with two windows to front in each pen, gives a guarantee there will be no lice or sickness in this house, while the roost beneath, not touching any outside wall of house, places the fowls where they will have pure air all night, and still no draft, as it will be enclosed with cloth on 3 sides and curtain for winter. A convenient dark nest is given where all fowls will have ample room and in cold weather the eggs will be kept warm by the other layers. The front scratching shed will have open wire all over the front except the lower part for 3 ft. high, and this can be closed by muslin window when it storms. The doors between front and back parts will be of muslin mostly, with a separate fowl door at bottom of each, and these doors will be used as the weather demands, and if a bunch of fowls can get into more congenial quarters for health and comfort of both fowls and attendant, then I have never seen them.

I have always studied the different styles of poultry houses that I have found on my trips, and while none of the ideas embodied in the plans I furnish here were ever copied from anyone else, still I claim they are the most practical and labor-saving of any poultry house construction I have ever seen, and all visitors say the same thing. The photographs show many points and I will mention a few advantages of each style of house for its particular purpose.

All poultry houses should be on ground sloping to the south, so as to be perfectly dry, and this construction guarantees the main floor to be dry, as it should be raised slightly, and if on a slope the front is also protected.

Perfect ventilation is afforded by bringing fresh air gradually through the muslin in lower front window, then through the muslin door between pens, up to peak of roof, where it escapes in muslin covered hole just under roof. No drafts are possible, and still perfect circulation is going on all the time. I am an advocate of muslin, but it is no good for windows alone, as it makes pens too dark, but by having small glass windows for light and muslin for air, you have it just right. In summer all can be thrown open, and you have cool quarters instead of a sweat box at night.

I recommend cement rat proof floors, and straw lofts supported by common wire netting, as this affords the most perfect ventilation and insures perfectly dry houses, which is the main point in any construction.

All partitions and fences between pens containing males should be at least 2 feet high, to prevent fighting.

Neat dusting box can be made by making a three-cornered box in corner of pen and filling it with road dust or sand. Provide hoppers for grit, oyster shells and charcoal.

While I have always recommended cement floors, still I have found them cold in winter, but recently I experimented with sawdust and discovered a way of correcting this. By making a mixture of sand and common sawdust from any sawmill (2 parts sand, 2 parts sawdust and one cement) you can make a cement that is just as hard as there is any need of, and to all outward appearance it is same as any cement floor, but it being half wood, it is not so cold.

The best dropping board I have ever used is made by making a form of the waste slabs from any sawmill, which can be bought cheapest of all lumber, and then by nailing smooth part down and rough uneven side up, you have the foundation to give a thin skim of Portland cement mixture, using cinders from mills for part if desired mixed with sand. This gives a smooth, hard floor that dropping can be removed at any time of year, while they often freeze to board floors, so as to be almost impossible to remove. This same style of house can be used with a flat floor even cheaper than the side hill slope, by placing roost in rear like old style.

DeGraff Movable Colony House.

The accompanying diagram shows the style of a movable colony house I have adopted, and while it is different from any I have ever seen, still for the cost and amount of desirable space received I consider it the best combination house I have ever seen. While the picture shows about everything, I will explain that two floors are furnished and the attendant can clean every inch of it very easily.

The roost is finished so it will be warm and air tight with drop muslin for winter. The eggs can be hunted from the outside of house, which saves a lot of time on such a large farm. The light from the two windows in front reaches every point and kills all disease. As the bottom hole at step is always left open, this will remove all impure air, and the muslin window at top of door just under the front part of roof gives the same results as a front muslin window and still does not cut off the light like a muslin window. This gives a continuous light circulation of air in and out of the house, at the front end, while the back part is warmer from being well made and air tight at all joints. It can be moved twice a month to new ground and in winter all these houses can be brought to a central point, where I expect to build a large rustic open shed for them to scratch and exercise in the winter, thereby affording as perfectly healthy conditions as I know how to design for this style of houses

NOTHING SUCCEEDS LIKE SUCCESS.

The Reds don't ask for any snap,
    They want no easy thing.
For trouble they don't care a rap;
    They take all you can bring.
They calculate to have to sweat—
    It's right, I guess, they should.
They don't care what knocks they get,
    If they can just make good—
               (Neither do I.)

MY FUTURE POLICY OF BUSINESS
Seven Good Reasons for Buying Reds Here

First. The Reds are unconditionally the greatest money making fowls on earth, and all amateurs in doubt about what breed to select can have no second choice.

Second. I have not only the best flock of Reds on earth, but they are raised and handled under the most ideal conditions, and they have a color so far superior to my competitors that they imagine they are kalsomined, but the best coloring is produced by nature under ideal conditions, as then a color is produced no words can describe or art reproduce.

Third. I have demonstrated beyond a doubt that I have the greatest prize-winning strain on record, and I will hereafter not exhibit any of my stock, except possibly at one show a year, and will supply my customers with guaranteed winners of a quality that no other breeder could afford to let go, as they have only one each of this kind, and they need them. As I will not weaken my best breeders by showing, you will get the benefit of the stronger vitality in all chicks and eggs sold.

Fourth. I have the selection of 5,000 Reds raised from my own stock and I can select any particular kind you may desire and give you better value than any small breeder could, and by giving you the benefit of my years of breeding, showing and advertising, you start right and do not have to lose time learning it all over yourself.

Fifth. I sell all birds on approval, guaranteeing satisfaction and delivery, and if on inspection they are not as represented, or satisfactory to you, I will refund your money on receipt of birds, less express. I am perfectly responsible for any order that may be entrusted to me.

Sixth. I have, through systematic advertising, built up a reputation known all over the world as "DeGraff's Vitality Reds"—"Best Reds in America." As I have the names and addresses of 50,000 admirers of Reds, I will not spend much more money advertising in papers, but will issue circulars to as many as I can get out, and list what I have to sell, and I am assured of a good business.

Seventh. My catalogues are the highest grade poultry literature ever published and contain colored pictures true to life, and information you can get nowhere else. My 1913 Book on Reds, with three colored pictures, far excels any of my past successes and contains a work of art showing ideal colors of Red males by using a new idea never before dreamed of.

CANADIAN BREEDERS, PLEASE LISTEN.

These words are in favor of a man, who, to my mind, has been too harshly treated by the powers that be. This man is E. T. DeGraff, of Amsterdam, N. Y.

Don't the readers of this Journal think it is about time he was given an opportunity to exercise the wonderful power he possesses on behalf of Reds?

To my mind, there has never been another breeder in the United States who has done as much for Reds as has DeGraff, and I know that any dealings I have ever had with this man have been perfectly satisfactory, which is more than I can say about some others, who at the present time are considered to be above reproach. And not only are the dealings I have had with him personally O. K., but in my position as President of the Canadian Rhode Island Red Club for three years, I have come across a great many others who have bought from him at some time or other, and without exception they say: "He always used me right." I have never met him personally; but from my knowledge of the chicken business I feel that he has not received a square deal, and would like to see him get it.
—S. Charlton, London, Ont., Canada.

THE CAUSE OF THE HIGH PRICES OF FRESH EGGS TODAY

In this day of phenomenal improvements we are accustomed to hear of startling developments in connection with wireless performances, the bird men, and the gasoline buggies, but where is there an older subject—and still more important today—than the prevailing price of strictly fresh eggs? This important part of the diet of all classes of people has been used as far back as any records can be found, and yet today, with all the highly advertised hatching and raising machines, together with the many new delusive "systems of making money from poultry," the price of eggs is, and will be, higher than ever before.

The poultry business is the biggest "little business" on earth, and Government statistics prove that the poultry industry is Uncle Sam's King Industry today when it is considered in all its many branches.

The number of eggs laid and consumed in this country is beyond comprehension; still there is always an unsatisfied demand for strictly fresh eggs, at the highest prices, all the year round. The development of the cold storage feature of the handling of eggs has changed conditions in many ways, still the real cause of this continually increasing price of eggs I blame to the parasite coccidia, which causes millions of young chicks to die annually, before two weeks old, no matter how well they may be cared for. This microbe is responsible for more failures in the poultry industry than any other one cause, and only recently has there been any light thrown on its origin, cause or cure, notwithstanding the best brains of the country have been investigating the cause for years.

All fowls should have free range, with liberal supply of green feed, pure water, pure air and clean, dry sleeping quarters, together with a variety of grain (whole and ground into a mixed mash), and when these conditions are complied with, very little sickness will ever come and a larger percent. of profit on the money invested can be secured from poultry than any stock on a farm or city lot.

There is a tendency to intensify poultry breeding by forcing nature in every way possible—by keeping too many fowls in small houses and small, dirty yards, hatching eggs laid by hens that have been forced to lay by stimulating foods and housed in warm houses, all of which reduces the vitality of chicks and fowls raised. I have seen it demonstrated where three generations of fowls raised under these conditions, and hatched in hot air incubators, could not perpetuate themselves, and were utterly worthless as a business proposition.

As there are over 100 varieties of standard bred poultry, and over half of the chickens raised are "Dunghills" or "Crosses," there naturally would be some breeds possessing qualities making them easier to raise than others. The unprecedented popularity of the Rhode Island Red fowls has been caused by the fact that these Reds are easier to raise than any other chicken, and when matured they are hardier, healthier, and therefore better winter layers, than many of the more delicate breeds.

While the unnatural conditions that fowls are kept under help to develop these germs, still it can be held under control by the proper use of Epsom salts in feed, scattering slacked lime over houses and grounds and disinfecting all eggs hatched, and using either hens or hot water incubators, instead of hot air machines, as there are many points about some hot air machines that help to develop this germ, and chicks taken from the machine in apparently healthy condition are destined to sure death, no matter how fed or brooded.

The price of eggs has advanced in proportion to the number of hot air incubators that are sold, showing the more nature is cheated the higher penalty we have to pay for the imposition. The amateur reads in the poultry papers of the wonderful results that can be accomplished by using the highly advertised machines for hatching and raising young chicks, and he dreams of the millions of dollars that can be made by increasing in proportion the wonderful results that can be secured from a small investment in a few hens and their chicks raised under ideal conditions, following nature's rules.

There is every reason to guard against all possible means of spreading disease, while the causative organism of the diseases of the respiratory and digestive tracts are passed out of the body of the birds in immense numbers in the droppings. These parasites are spread all over your own grounds and often given to or received from your neighbors, where the disease has been allowed to go unchecked. Even the wind and sparrows act as agents to spread these deadly germs, especially in thickly settled communities.

Any pathologist will tell you there are many intestinal parasites, be they worms, molds, bacteria, or protozoa, that may produce disease or develop conditions in which certain well known disease-producing parasites may operate.

All eggs must pass through the cloaca, which gives passage likewise to the droppings, thereby giving an opportunity for all infestious microbes of intestines to contaminate the shell of the egg unseen to the naked eye. All experts in artificial hatching realize the importance of washing all eggs before placing same in incubators, as any parasite that might be attached to the shell would be developed by heat of incubation and remain lying in wait for the young chicks when hatched. As the young chicks are in the machine for over 24 hours at least, they are almost sure to be infected by these germs, and a large proportion are destined to die, and no power yet discovered can save them. The amateur breeder does not realize the true cause and blames it all in the feed, brooder or dealer who sold him the eggs, while the true cause in the incubator goes unblamed, as the chicks were apparently healthy when taken from it.

While eggs from perfectly healthy hens are free from disease, still there may be some eggs in large machines that were laid by hens that are parasitized with coccidia, and the chicks hatched will all be affected in proportion to the vitality of the individual specimen, and even though the chicks may grow for some months, they will never make healthy, well-developed birds.

This long-dreaded disease of young chicks, commonly called "White Diarrhoea," has been the cause of thousands of dollars being expended and years of experimenting by the best talent in the poultry business, with no apparent result, as results vary so much under different conditions. A positive cure would be a Godsend to the poultry breeder, and even the following pointers will be of great benefit to all persons that have ever suffered from this scourge:

Hatch eggs from only the healthiest of fowls, and have these on free range with plenty of green feed. Hatch eggs under hens set on the ground, or in hot water incubators, and brood with hens or hot water brooders, with plenty of fresh air and as low temperature as they will stand without crowding. Give young chicks free range and house in open air colony houses. Feed all laying fowls an occasional feed of Epsom salts in wet mash by dissolving one-third of a teaspoonful, to each fowl, in water, and then use this water to make the mash, stirring thoroughly. Purify all drinking water by using enough permanganate of potash to color the water. Scatter air-slacked lime through houses and yards, especially the dropping boards, and keep litter free from any mold. Disinfect all eggs by washing in 90% of alcohol and thoroughly clean and disinfect all incubators and brooders. If these conditions are carried out you will receive far better results than ever before.

After chicks are hatched, feed nothing but sour milk for two or three days, and continue its use as long as you can, and your chicks will mature all the better for it.

The acid in sour milk has a tendency to kill this germ and prevent contagion better than anything yet discovered. My drinking bottles are especially adapted to feed this kind of milk, and the iron rust from base is another natural remedy that benefits the health of chicks.

Naturally everyone has their preference in color and style of a fowl they admire, but after 20 years of breeding of 20 varieties of fowls I would advise all new beginners to start with some recognized breed, and in my estimation there is no second choice in comparison with the Rhode Island Red, when properly bred.

Millions of families are so situated that they could, at a very small expense, provide a cheap shelter (and the fresh air plan is best) to house a dozen hens, and these would almost keep the family in eggs from the table scraps that are thrown away from the average family. Where is there any really appetizing dish that does not have eggs used in connection with it in some way, and where is there any dish that would not be benefited by having more strictly fresh eggs in it if we could but buy them at a more reasonable price?

While the prevailing price of eating eggs seems high from this standpoint, still I am selling every egg my hens lay, for hatching, the year round at 10 cents each for the lowest price, and often $1 and $2 each are paid for individual pedigree eggs for hatching prize winners.

A VICTIM OF TUBERCULOSIS

The accompanying picture shows the familiar kind you have all seen, and the sooner this kind is killed the better you are off, as they will never be worth the food they eat.

THOROUGHBRED ST. BERNARD PUPS FOR SALE

After trying several breeds of dogs for watch dogs, I have decided the St. Bernard is the best all-round dog for any poultryman's use. I find the size of the St. Bernard is sufficient to keep the average trespasser away, while if necessary to use force, the St. Bernard can be relied on every time, and still they are very quiet and docile. They are excellent protectors for children, and I would not part with my stock of St. Bernards for a lot of money.

I have them for protection in orchard and can furnish a limited number of pups at reasonable prices at certain seasons of the year.

HIGH GRADE GUERNSEY CALVES FOR SALE.

I consider the Guernsey cattle are in their class what the Reds are in fowls, namely, the best of their kind.

I keep a herd of thoroughbred Guernseys for butter and use the surplus milk for chicks and fowls, and I find there is no other feed that takes its place at all ages of fowls. My barn, 70 feet by 40 feet, has all the modern improvements I have ever heard of and "then some more" of my own invention.

Cement floors have always been pronounced cold and slippery, so I made my floors of half common sawdust, together with best of sand and cement, and I find it is not as cold and not as slippery and proves very satisfactory. I have an air space of over six inches under the part that the cows lay on, which is connected with a galvanized shaft running outside of barn to roof. An automatic blower on top of shaft draws the bad air continually through small openings back of each cow, connecting with the manure trough, thereby affording perfect ventilation without any possible draft. To my knowledge this is the first time this has ever been done, and every visitor tells me it is the most modern stable they have ever seen. Iron stanchions and running water, with vitrified silo, make the equipment about as near perfect 100 point stable as any health office could demand.

PRIVATE CAMPAIGN OF PUBLICITY
(Editorial in Red Breeders' Bulletin)

When Edward T. DeGraff of Amsterdam, N. Y., thought of photographing the most important commercial ages of the Rhode Island Red fowls, so as to show their many good points, he did the breed an everlasting benefit, that has been a benefit to every breeder of this popular variety. His article on Reds, showing their many good utility points and illustrated by a set of ten different ages of the breed, has been published in every poultry paper of any account in this country, besides being published in leading papers in England and Germany. It was translated into Spanish and went to every Spanish speaking country of the globe, thereby introducing it to thousands of readers that had never heard of this new breed which is fast proving itself to be the Greatest Utility Fowl on Earth.

DeGraff's Color Plate Book on Reds has filled a long felt want, and it shows it is appreciated by the Red fanciers, as his issue of 5,000 copies was sold at 25 cents each before the year was out.

Realizing that a color plate of the breed was in great demand and that the old club had decided it could not afford the expense of getting out such a picture, he has tried at great expense, by different methods, to reproduce this ideal color for the benefit of the breed. His 1907 picture of a pen of Reds in color was a great success, and his 1908 embossed Red cock bird in true colors was a grander success, but his 1909 picture is a wonder, and the first picture ever taken of fowls by this process in the world. The four-color process plates are reproduced direct from the fowls themselves, which will guarantee an exact reproduction of color, and the subjects which were shown at the Detroit show, as the birds that will make pictures, were the selection of over 10,000 Reds handled by Judge DeGraff, and embody both the ideal shade of red and perfect harmony in all sections.

WELL BRED DAY OLD CHICKS $25 PER 100.

A limited number of Day Old Chicks at 25c. each in any number over 30, if ordered 3 weeks in advance, so I can have them hatched for you when wanted. I advise you to buy eggs, but if you want to try the chicks I can deliver you healthy, strong, farm-raised stock that will live.

I consider the H. & D. cold-air brooder the best of its kind, and I will ship Day Old Chicks in this arrangement for $2 extra, the actual cost of same.

Fully Developed Cock   Matured Yearling Cock   Half-grown Cockerel

MY PRACTICAL ILLUSTRATED INTERPRETATION OF THE AMERICAN STANDARD OF PERFECTION, AS APPLIED TO RHODE ISLAND RED FOWLS.

Judging by the general tone of hundreds of letters I get from amateurs, and the questions I have heard asked at the large shows I have attended this winter, there is a wonderful variation as to how the Standard is practically interpreted.

As I have studied the judging of Judges Riggs, Bryant, Card and Weed at the four largest shows held this winter, and taken photographs of most of the winners, for my reports of these shows in POULTRY, I feel I have had the benefit of an experience that few have taken the trouble to follow up. I intend to give the reader the benefit of what I have learned, from 10 years practical experience in breeding and exhibiting Reds, and my careful study necessary to develop my true to life colored pictures which are acknowledged to be the highest grade poultry pictures ever published.

Many look upon the Standard as some sacred epistle that has been handed down to us poor mortals here below, and don't seem to know the why and wherefore of its existence at all.

The American Standard of Perfection printed and copyrighted by the A. P. A. is nothing but a compromise of the many different opinions that may have prevailed, as to what the ideal of any breed of fowls should be, to govern its future development, and establish a foundation for judges, to base their decisions, in selecting the prize winners at our poultry shows.

At Falls River, Mass., in 1898, there was held a meeting of few but enthusiastic breeders of Reds, who formed a club, and voted the officers there elected the power to act in any way to promote the interests of the breed. In 1901 a meeting was held at Boston at which it was decided to adopt and copyright a Standard, on the lines of what had generally been the accepted ideal, after much discussion pro and con. At the A. P. A. meeting held at Rochester, 1903, several prominent Eastern Club members combined their best birds and made a very creditable showing of what had up to this time been christened by the old line breeders, "The Great American Dunghill."

Largely through the efforts of Hon. C. M. Bryant, who had acted as President of the Club for several years, the S. C. Rhode Island Reds were admitted to the Standard, placing them on equal terms with all other breeds, in the race for public approval.

Unfortunately there was so much wire pulling, and too many interests to be satisfied, that sufficient forethought was not exercised among the officials of the A. P. A., that the American Reds and Buckeye Reds were also admitted at the same time.

This action aroused strong sentiment in the Club, and there was much public discussion for a year afterwards, until the American Reds were wiped off the face of the Earth and Rose Comb Rhode Island Reds were admitted to the Standard as a second variety of the Red Breed. That it was a mistake to ever have admitted the Buckeyes as a breed has been proven, as they evidently will never win public approval, and I hope this experience will be realized if the Rhode Island Whites ever attempt to be admitted as a breed, as there can be no such a breed, distinct from other white fowls.

When the Standard pictures were adopted, it was very noticeable that the ideal profiles and the printed description did not jibe, and it was evident at some shows that the judge's opinion did not jibe with either one, as he was judging entirely by what he thought should win. The popularity of the breed grew so rapidly, that naturally there arose jealousy and discontent among the breeders, because the Club officials were always confined to a select clique in and around Boston and Fall River, so that several opposition Red Clubs were formed, and the accepted standard received much discussion, part of which was adopted at the revision meeting held at Niagara Falls in 1909. To show how fickle public opinion is, I will state that the A. P. A. revision committee when it met at Chicago, Ill., adopted all the suggestions made by the Western Red Club, and when the Eastern Red Club made a united effort, they succeeded in changing all the points they desired, except a few, and the present standard was adopted, which cannot be changed till 1915.

The principal point of contention among breeders was whether there should be ticking in hackles of females or not, and whether it should be allowed in males also, which was decided to go against nature and have ticking in hackles of females, and not in males, so that every show held in the country today, we see males shown with a large part of their hackles pulled out to remove the ticking, which the judges overlooking so far, as an example of tampering with nature, while males of all ages are

Fully Developed Hen   Matured Yearling Hen   Half-grown Pullet

being sold, with the continual uncertainty, whether the young growing cockerel will lose his ticking or not, and whether the cockerel that is free from it, as a cockerel may not develop it when he moults, all of which is very unsatisfactory to the amateur breeder, who feels he is being humbugged.

As the reading matter and profile cuts of Standard are copyrighted, and you cannot buy any separate breed standards for Reds, you will have to buy the whole book to learn what you want to know about Reds, but I have the privilege of giving you my interpretation of what I consider ideal specimens by showing pictures from life of birds that I consider as near my ideal as any I have ever seen. The motive of the standard is to give the amateur information that will enable him to know his prize-winning Reds from his inferior stock, but when we consider that the same cuts are used and same description applied to old and young of all ages, you can see the object is not entirely accomplished, and much individual interpretation is exercised in its practical demonstration. As competition in shows include all ages, I will show the differences that take place, as the bird matures, although equally high grade specimen at each age. The promising young cockerel shown just before reaching standard weight, gives promise in every way of being an exceptionally fine specimen, still the judge should not bank too much on promises, as very often they are not fulfilled. The finishing touches of any bird may develop faults that could not be foreseen, such as white in under color, white feathers, ticking in hackle, may come and may disappear, while some males change the angle of their tail carriage entirely, and even change their type entirely in a month's growth. The matured cockerel shows a very well proportioned, well balanced male just as he reached Standard weight, and there is little doubt but he will develop into a grand good cock bird, and may look much like the type shown in the matured cock picture, as this shows a male fully matured, rounded out in all parts, with full sickled tail, which seems to be a weak point in so many really good Reds in other respects. The oblong body is very marked in each specimen, and you must always make some allowance in the apparent length of legs of photographs taken at this elevation, as we are accustomed to look down at a bird. While the Standard color, "rich even red," applies to pullets, hens and old hens, still I have seen but two hens this whole winter's showing, that could be said to be anywhere near the color we are giving prizes to in pullets, and one of these, "Lady Graffmoore," took the color special over all pullets in show, which proves her color was what we are all looking for, but seldom ever seen in hens of any age.

The amateur should know that mighty few Red Hens ever have the color they had as pullets, but will moult out lighter, and if of a good even harmony, although lighter, you have what any judge will consider above the average hen, if her type is right. It is a mistake to make the amateur believe that anyone has hens, he can buy of the Standard color, by claiming non-fading, as while visitors tell me I have the reddest flock of hens they have ever seen as a flock, and I have twice delivered from my stock First Prize hens at N. Y. Shows that have won Club color specials over pullets shown, still I will admit I cannot sell you hens with the color of prize pullets, no matter what you pay for them.

The pullet picture shows that velvety finish that often decides the winner in pullet classes, which I feel is sometimes a mistake, as this bloom of perfect health may disappear more or less as she lays eggs, while the typical formed pullet will not lose, but improve, as she gets busy, so shape should count half of any prize bird, and for utility purposes type counts much more. The hen picture shows one of best balanced hens that I have photographed this winter, and she is as good as the picture throughout. The old hen when taken was 7 years old, so you can excuse the unevenness of plumage, but her vitality shows out strong for her age. The other hen picture was taken 8 years ago, and I have never taken a picture since that shows the strong desirable points of a red hen better than this, namely, vitality, ruggedness, alertness, wide breast, low, wide tail well spread, and ability to forage and scratch for her living if circumstances are such she can do it.

The Faded Old Hen   The Foraging Type

Photo of Ideal Wing

My color plates give the ideal shade of color of each sex better than words can describe, and my spread wing picture has been pronounced, even by my competitors, the most educational picture ever printed in poultry industry, which gives me much satisfaction, especially after the Red color pictures attempted by A. P. A. have proven such flat failures, mostly because the idea was not developed in a practical, common sense way.

While the Standard description of what color a wing should be, generally gets the amateur all at sea, and the color of the small feathers at base of wing is not given at all, still I believe the accompanying pictures will give the idea where and how much black there should be on wings of both male and female, while females that get it as strong as this are very scarce, and if right otherwise, very valuable, either for showing or breeding.

Ideal Rose and Single Comb Heads

When we stop and think of the wonderful improvement that has been made in Reds, in this short time, and their wonderful unprecedented popularity ever increasing, we should realize that we have a grand opportunity in them for future development. As we pass through this world but once, any good therefore that we can do or any kindness we can show to any human being, let us do it now, lest we defer or neglect it, for we shall not pass this way again.

If your Reds have not been coming as you would just like to have them, remember they are made up of crosses for years back, and are liable to strike back to inferior ancestors, and try to remedy your weakness by buying the best you can buy to correct it.

THE POULTRY HOBBY.

There is no denying it, every mother's son of us has some hobby or bug on which he is not exactly right mentally, and they are generally cranks on this particular subject, and in most cases know what they are talking about, and the true value of the true quality in that particular line. Of all hobbies, I know of no more pardonable one than the poultry hobby, as there is a possibility of its paying its own way, while most hobbies are money sinkers and no satisfaction or benefit to anyone. America is a wonderful country and we are a wonderful people, and the larger my mail gets the more I realize what a wonderful thing human nature is anyway. All nations of the earth look up to the United States, and their citizens look forward to coming to this country as a sort of heaven on earth. It certainly is a wonderful country to live in and the mail order business possibilities open to the successful advertiser is beyond the imagination. This land of the free—land of fine churches and 40,060 licensed saloons; Bibles, forts and guns, houses of prostitution; millionaires and paupers; theologians and thieves; liberists and liars; politicians and poverty; Christians and chain gangs; schools and scalawags; trusts and tramps; money and misery; homes and hunger; virtue and vice; a land where you can get a good Bible for fifteen cents and a bad drink of whiskey for five cents; where we have a man in congress with three wives and a lot in the penitentiary for having two wives; where some men make sausage out of their wives, and some want to eat them raw; where we make bologna out of dogs, canned beef out of horses and sick cows, and corpses out of the people who eat it; where we put a man in jail for not having the means of support and on the rock pile for asking for a job of work; where we license bawdy houses and fine men for preaching Christ on the street corners; where we have a congress of 400 men who make laws, and a supreme court of nine men who set them aside; where good whiskey makes bad men and bad men make good whiskey; where newspapers are paid for suppressing the truth and made rich for teaching a lie; where professors draw their convictions from the same place they do their salaries; where preachers are paid $25,000 a year to dodge the devil and tickle the ears of the wealthy; where business consists of getting hold of property in any way that won't land you in the penitentiary; where trusts "hold up" and poverty "holds down"; where men vote for what they do not want for fear they won't get what they do want by voting for it; where "niggers" can vote and women can't; where a girl who goes wrong is made an outcast and her male partner flourishes as a gentleman; where the political wire-puller has displaced the patriotic statesman; where we have prayers on the floor of our national capitol and whiskey in the cellar; where to be virtuous is to be lonesome and to be honest is to be a crank; where we pay $15,000 for a dog and fifteen cents a dozen to a poor woman for making shirts; where we teach the "untutored" Indian eternal life from the Bible and kill him off with bad whiskey; where we put a man in jail for stealing a loaf of bread and in congress for stealing a railroad; where the check-book talks, sin walks in broad daylight, justice is asleep, crime runs amuck, corruption permeates our whole social and political fabric.

While in the poultry world, stranger things than these are happening every day. Incubators are being sold that will not hatch. Brooders are sold that will not brood, advertisements are being placed that will not sell, but if you exchange a bird with your friend for a show, and admit it, you are liable to expulsion for life from all privileges of the American Protective Association of Fakirs for admitting it.

THE SUN NEVER SETS ON DeGRAFF'S REDS.

After several years' experience in shipping fowls to nearly every civilized country of the globe I feel safe in guaranteeing delivery to almost any place on earth, as by actual experience I have never had a bird die on an export order. I have designed an original shipping coop which is very light and still as solid as a rock, and contains all necessary apparatus to properly care for fowls on the way, including water dish and feed hopper, and I give ample room so fowls arrive in perfect condition. I understand the necessary export routine that has to be gone through and I have made arrangements with the American Express Co. to forward my shipments with charges following, and same to be at lowest rates possible. I recommend sending your money by international money order (not inland orders) or by American Express Co. order, forwarded in registered letter, and then you will get your order filled promptly without delay or correspondence, as I ship to some places where it takes two months to get an answer to letters.

I have a Spanish circular which I send with all letters desiring it, and I have an interpreter of all languages, so we are in shape to handle all kinds of export orders, which is fast getting to be an important part of my business. Any persons desiring to purchase any other American breeds of fowls I will be glad to act as their agents and ship only first-class stock at reasonable prices along with any of my breeds or alone.

Original Style of Export Coop, Used for All Foreign Shipments, Light and Strong

MY RED MISSIONARY WORK FOR BETTER POULTRY IN ALL PARTS OF THE EARTH.

My articles on Reds printed in nearly every language has attracted attention to the many good qualities of the Reds, and I receive letters every day inquiring about and ordering Reds from some new country, and invariably I have delivered the birds in perfect health and the birds have demonstrated all I claim for them. I have letters from France claiming the Reds would outlay any continental breed there, and Germany and Prussia, are fast importing foundation stock, as the breed has quite a hold already in those countries.

At the recent Club Show at York, England, where there was a class of 85 Reds, the 1st and 2nd prize males, and 1st and 2nd prize females were Reds I sold, or bred from birds I sold, so you see my stock has quite a start over there, and the Club is fast growing in membership. I have a letter from Alaska saying that a pen I sent there had averaged 192 eggs each for the year, and ordered more.

Good reports come from Australia, and Hawaii Islands, while all the countries of South America, are inquiring and ordering a few to give them a trial, and it is wonderful how this breed is adapted to both warm and cold countries, as good reports come from both in every case.

DeGraff's Reds Always Win the Cups.

SEVEN SILVER CUPS AND $100 CAPTURED BY LOCAL BREEDER AT SCHENECTADY POULTRY SHOW.

The prize record of the DeGraff poultry farm of Amsterdam, N. Y., at this show is something to be proud of, and it proves conclusively that their stock is all they advertise, "The Best Reds in America."

When we consider that the Rhode Island Reds are the most popular fowls of the day, and the Red class was the largest in the show, and the show of Reds was the largest and best ever held in the state and outclassing both the New York and Boston shows in number and quality, the following record is remarkable:

S. C. R. I.—1, 2, 3, cocks; 1, 2, 3, hens; 1, 3, 4, 5, cockerels; 1, 2, 3, pullets; 1, 2, 3, pen old; 1, 2, 4 pen young.

R. C. R. I. Reds—1 and 2, cock; 1 and 5, hen; 2 and 3, cockerel; 1 and 3, pullets; 1, pen old; 1, pen young.

Besides the $25 gold sweepstakes special for the largest and best exhibit of the show, and the $25 sweepstake special for the largest and best exhibit of Reds.

Both New York state silver cups for the largest and best exhibits of both breeds of Reds, offered by the Rhode Island Red Club of America. Both cups and silver salad bowl, offered by the National Single Comb Rhode Island Red Club of America, for the best pen of old and best pen of young and the ten highest scoring birds.

Lester Tompkins took two days to judge the Reds alone, but his awards gaves perfect satisfaction.—Schenectady Gazette.

IT IS BETTER TO START WITH ONE GOOD PAIR THAN MORE INFERIOR SPECIMENS

MORE FOWL POETRY

Said the old Red rooster,
　To the old Red hen,
You haven't stopped laying
　Since the Lord knows when.

Said the old Red hen
　To the old Red rooster,
I've been laying so long
　I have kinder got used-ter.

MY 500-FOOT WINTER LAYING HOUSE NOW UNDER CONSTRUCTION

PRACTICAL, ECONOMICAL, ORIGINAL IDEAS.
(From Commercial Poultry, Marceilles, Ill.)

Originality is a gift, and it cannot be learned or taught in our schools or colleges, but must be born in the individual, to be the genuine article. It is just as natural for some to do things in a new and original way as it is for the vast majority to never think of departing from the antiquated ways their forefathers have followed for centuries.

At Amsterdam, N. Y., there is a poultry farm conducted by Edward T. DeGraff, who has many original labor-saving ideas in operation that could be followed to the advantage of all poultry raisers.

Anyone contemplating going into the poultry business on a large scale would be well repaid for a visit to this farm, as it is located on the N. Y. C. & H. R. R. R. and the interurban electric road from Albany to Gloversville, N. Y., has a private stop at Teller's, right at the yards, and visitors are always welcome to inspect the houses and stock.

Accompanying cut explains his watering system, which run the year round and furnishes pens with all the fresh running spring water that they care to drink; and as a very large per cent. of the egg is water, this is very important, not alone from the labor saved, but the better quality of the eggs laid and the general health of the fowls—as more disease comes from impure water than from any one cause.

He raises his chicks, in large numbers, by using indoor brooder in his piano box colony houses, which are both entirely of his own design and furnishes each pen of chicks with a separate dish of fresh spring water by this ingenious device, thereby saving a lot of work and giving the young chicks pure water. He runs a supply pipe to a half barrel in the field and to the bottom of this barrel with a shut-off to it is connected a ¾-inch pipe, which runs to all parts of the field, and at each place where a dish of water is needed, a coupling with a small hole bored through one side so that it remains on the upper side over the drinking dish.

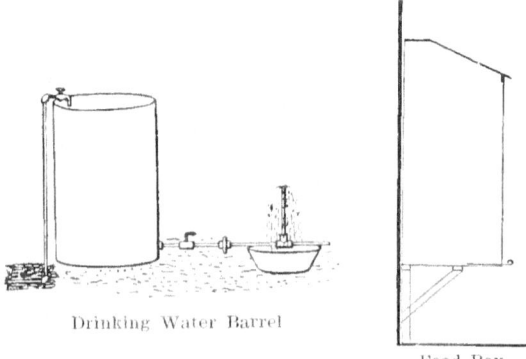

Drinking Water Barrel

Feed Box

Filling the barrel and opening shut-off of the pipe carries the water to all dishes, and as the combined capacity of all small holes does not equal the size of the pipe all dishes are filled at the same time. The field is filled with miniature geysers and can be repeated as often as desired each day.

The large oblong shoe boxes which are about four feet long make a very handy feed box to keep grain in, as they will hold more than a barrel and take no more room on the floor, and if a row of them are arranged on the most exposed side of your house you will have a partition that no cold can get through. Make the outlet hole in the lower end and stand about eighteen inches above the floor. Close the lower hole by sliding sheet of galvanized iron which slides inside the box through a slit in the front of box. When you want your feed, all you have to do is to put your pail under the box and draw the slide and get all the feed desired. The only hard part of the job is to keep the feed box from getting empty.

Meat scrap should be one of the most important things in all fowl's rations, and as it is quite expensive it must be fed in a way that it will not be wasted, and after experimenting with many kinds of hoppers he has decided the herring kegs beat them all. Place a round stone in bottom of keg, fill half full with sand and gravel, then put in about a quart of meat scrap and they will not waste a particle and you can always see when they need more, and the more gravel they eat the better for them.

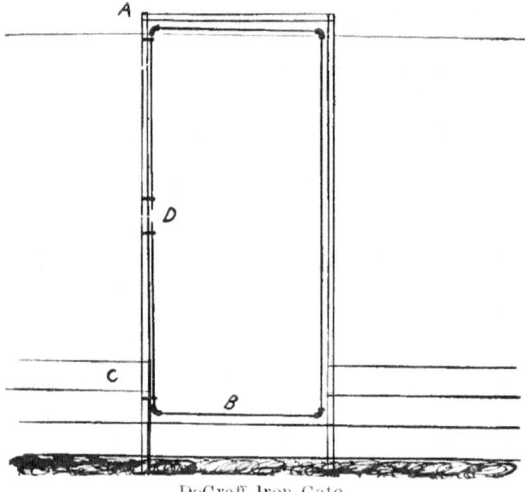

DeGraff Iron Gate.

A very neat gate to go with wire fencing is made from cast-off piping, which can be bought at any junk shop at less than a cent a pound, and by using four "L's" you can make a frame, and by using four large staples to swing on, you have a gate that will outlast the rest of the fence and will never sag. When putting up the wire continue it right over the gate and then twist the loose end around front part of gate till you have a neat finished job. Keep gate closed with pulley and weight and your fowls will never get mixed in the breeding.

Large shipping coops can be made from Shredded Wheat, Maple Flake, Force and other smaller sized boxes that have extremely light wood in their construction, making as light a box as can be bought at any price.

By melting the top off canned goods cans carefully you have a fine drinking dish for shipping fowls.

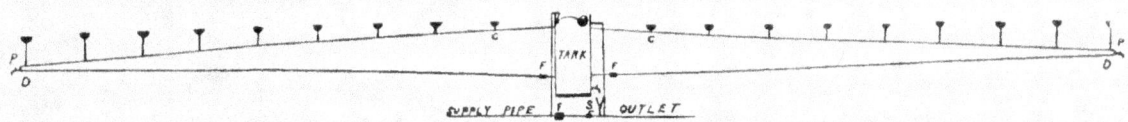

DeGRAFF WARM WATER SYSTEM.

Above cut represents a system of running water that can be used all the year round, by supplying heat in some manner to the tank shown, either by a large lamp under or by connection with a small hot water stove as I will have it.

As the water level of all cups and tank is the same any number of cups can be filled and kept just so full if properly installed, and by opening shut-off pipes will be emptied at once, removing all filth and preventing all chances of freezing in winter.

The cut explains all points necessary to install, but for explanation I will say that the supply of water is regulated in winter by the floating ball in tank for winter and left open in summer if supply of water is large enough. The two extreme ends should be supplied with a faucet, D, and plug, P, for purpose of drawing water when needed and removing any stoppage, if any. Unions should be placed where designated by F, for convenience of cleaning if desired. The supply and outlet pipes can be kept below frost line, and the one shut-off S, will operate the whole system.

I can supply the cups if anyone wants to install the system, and give any information not understood from above.

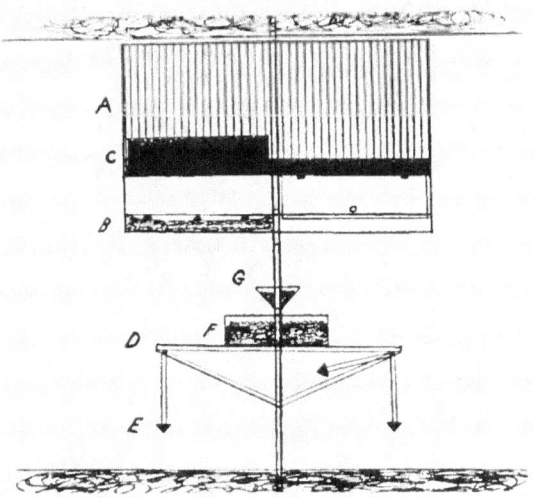

A, coops for surplus cockerels or setting hens. B, nests closed by hinged door. C-D, platform for drinking dish and hopper for both pens. E, swinging feed trough, which can be easily fastened back. F, on the level hopper with wire screen on feed. G, place for grit, oyster shells and charcoal. H, straw loft above coops for ventilation.

The accompanying cut shows how a partition between pens can be utilized to good advantage and still not take any floor space whatever, and every idea is a labor saver to the attendant.

The burlap that comes around coffins can be made very useful in many ways in hen houses, and it is generally given away. This can be used for fresh air windows and by stretching along wire partitions you make a division that males will not fight through and still give plenty of air and be very cheap.

The barrels of cork that Malaga grapes come in can generally be got for the asking, and the barrels come very handy for nests and the cork is the best material possible for packing eggs for shipment.

During the summer you can get the best of egg cases for 5c. each, and they can be used later when cases are scarce for eggs, or the wood can be used in many ways to make light boxes for shipping smaller orders of eggs.

A simple lice proof roost can be made from two scantlings cut at the desired length and planed smooth on all sides and rounded on top. Get some old or new broom handles and bore eight holes just the size of the handles, and then you can make a perfectly tight joint that no louse can hide in. Bore four holes on lower side for legs, which should be just high enough so fowls will not try to crawl under them, then four holes between sticks for braces which will make a perfectly solid frame that will stand on its own legs on the dropping board and be easily removed to clean roost, and every inch of it can be seen and painted with lice paint. This arrangement leaves no connection with the sides or ends, which is always the runway for the lice and mites to the side of house, where they hide during the day. This roost, combined with a dusting box made from the common sized soap box, filled with sifted coal ashes (often renewed), keeps fowls perfectly free from these little enemies which are often responsible for the remarks from new beginners that poultry does not pay.

He never nails anything to the side of the house so that it cannot be removed when he whitewashes. This can be accomplished by driving the nails part way into the boards and then boring small holes in box so that they can be placed right over the nails.

As lumber is quite expensive now he gathers up the old shipping boxes which grocery and shoe stores have to sell at from five to fifteen cents each, and by using the right size, you not only get cheap lumber, but you save the time of making the box.

A nest possessing many advantages can be made from the Shredded Wheat boxes by removing one end and making the top into a slanting cover so the fowls cannot roost on it. Hang it on the wall with the open end away from the light and you have a dark nest in which four hens can lay at once and still eggs can be easily hunted by raising the cover. This style of nest is a great help towards preventing chilling of the eggs when they are wanted for hatching early in the season. As during the forenoon there would generally be one hen on the nest all the time, and she would cover all eggs laid, and by hunting about noon you would have eggs that would hatch, provided the fowls had the proper care and plenty of pure, fresh air. Make nest material out of chopped tobacco stems and no lice will stay.

I PRACTICE WHAT I PREACH

"If the day looks kinder gloomy
    An' your chances kinder slim,
If the situation's puzzlin'
    An' the prospects awful grim,
An' perplexities keep pressin'
    Till all hope is nearly gone,
Jus' bristle up and grit your teeth,
    An' keep on keepin' on."

(That's how I got my start.)

My Famous S. C. Red, "AMSTERDAM,"
Photo of my ideal type of cockerel.

THE LEADING RED SPECIALIST OF AMERICA.

Many do not realize the advantage of buying from a specialist in their respective lines of trade, but if you will only stop and think of the many advantages gained by so doing, you will trade with no other. A specialist is supposed to know all about one thing. For instance, what chance has the small breeder to fill your order along side of my facilities for doing business in my line? How many male birds can any small breeder show that are really right, while I have on hand over 100 breeding males that are strictly high grade all the year round and sometimes I have 300 males for filling orders, and can select most any particular kind of a breeder you may need. While the small breeder may have one good breeding pen and needs most of those eggs for his own hatching, I have 50 breeding yards that are housed separately and every pen headed by a male worth more than $25 and a grand breeder for the pen he is selected for, so that I can fill all orders promptly in any quantity or quality and give good value for money sent. I consider eggs laid by the pick of 10,000 Reds sired by the best male I have ever seen worth $2 each to me or anyone that wants to get one of the very best Reds possible to secure. I sold $1,000 worth of eggs from my great sire "Amsterdam" at $1 each, and seldom had a kick on results. While the pens headed by prize winners prove they must have quality behind them by their winning in strongest shows of the country, still some of the males in my Standard Pens at $5 for 15 eggs are as good as the best that other smaller breeders have in their best pens, and ask high prices for. I have had so much trouble with the Express people breaking eggs that I have left nothing undone that can render the eggs unbreakable, until now I believe I have them packed so they can't break them, and I know of no other breeder that packs them as I do. I breed not only for the ideal red color, but I am very particular about type of my breeders, which many breeders seem to overlook. In the fall of the year I can give extra good value in breeding males, as I sell all my surplus breeding cocks and have half grown cockerels that show great promise at comparatively low prices. Remember my advertising pays for itself, and you get the value paid, in the eggs or fowls bought, as I don't have to charge up 50% to advertising, to keep up the false public sentiment fostered by the poultry press, just to catch the amateur.

There is no doubt but what the Rhode Island Red fowls have only begun to enjoy the boom that is bound to increase until they are known all the world over as the greatest utility fowl on Earth, and the honest element of the fancy of all acknowledge now that I have done more for the breed than anyone, or in fact, the whole club put together. This is all the reward I ask, as where is there any greater satisfaction than to be of some service to your fellow man, instead of continually hammering every other opposition breeder, like the most of the so-called fanciers do. My office has all the modern equipment for transacting all kinds of business to advantage, and all the leading poultry papers of the world are on file here. The walls of the room are covered with hundreds of prize ribbons and no end of silver cups, together with mounted ideal specimens of the breed, help to make a very artistic room. My shipping room has all modern equipment of a carpenter and plumbing shop combined, and we are able to do all this work ourselves; in fact, nearly all buildings have been built by students.

After experimenting with about every known method of brooding young chickens, I am now constructing an entirely new method of rearing chicks that will be different from anything ever used to my knowledge. The chicks will have all the advantages of nature, and the improvements that latest inventions can secure, and they can be cared for in any number desired up to thousands. The same idea can be adapted to incubation, and as I am having it patented I will soon give it out in poultry papers.

My catalogues have been so much appreciated that I hope soon to make this issue twice a year, and write an article each month for "Poultry" as I know a strictly unbiased poultry paper that is run for the benefit of the reader instead of the advertiser and will stand up for fair play to all, will be greatly appreciated.

My Famous R. C. Red, "SUCCESS,"
Photo of my ideal type of cock.

DeGRAFF'S RED BREEDERS' CLAMBAKE, SEPTEMBER 4, 1910

MY ANNUAL RED BREEDERS' CLAMBAKE, Sept. 1 (Labor Day), 1913.

For many years I have entertained the breeders of this locality with an annual clambake given in my poultry orchard, and it has grown to be of such importance that breeders have come from several states. My 1913 clambake will be held Labor Day, Sept. 1, 1913, and I hereby cordially invite every poultry fancier on earth (except Tracey) to meet with us on this day and enjoy a good feed among good fellows, and, incidentally, see for yourself the finest flock of Reds on earth, housed and cared for under the most ideal conditions you could ever imagine.

All I ask is your acceptance of invitation that I may provide for you. I will prove to you that I have, and do sell, eggs from over fifty breeding pens of strictly first-class Reds, all housed and yarded separately in large yards, and that they, together with my free range stock, make up a flock of thousands that can honestly be called Reds that are red.

I HAVE THOUSANDS OF REDS TO SELECT FROM

In order to do the amount of business I am doing in both stock and eggs it is necessary for me to have the selection of a large number of birds raised each year. Besides my own farm, where I raise and house over 2,000 each year, I have five branch farms each raising about 500 head each year under the most favorable circumstances, direct from my stock and under my supervision, and I often find these free range farm raised stock the superior of my own raised under different conditions. I will be able to supply **guaranteed winners** for all fall shows, as I have a fine stock of young ones already on the way that promise great results both from their breeding and present appearance.

> At Amsterdam, New York State,
> DeGraff's Best Reds are up-to-date.
> He breeds them pure for eggs and feather.
> With constitutions as tough as leather.
> —David Robertson, Easton, Md.

HALF GROWN FOWLS CHEAPER IN FALL.

While I sell a great many day old chicks I do not care to sell any after that until at least 4 months old as a partly developed chick can not be determined what it will really be when matured, but after that age I can sell young birds at much lower prices than when fully matured and often you get birds that will develop into prize winners that triple the price would not have bought.

I am always proud to show visitors my young stock, as they always remark that they have never seen a flock of Reds that were red before and it helps to convince them that there is a great difference what strain of Reds they start with for their foundation stock.

THE MAN THAT CARRIES THE FEED PAIL MAKES OR BREAKS THE POULTRY BUSINESS MANY A TIME

After suffering for several years, the lack of a strictly capable practical poultry man to care for my Reds, I have at last secured the assistance of one as good as I could ever expect to get.

Six years ago Mr. Earl Kidd came from Indiana and took a student's course for six months here at this farm, after which he took a winter course at Cornell Poultry College, Ithica, N. Y., receiving his diploma.

He put in one year on Candee Incubator Farm and two years as Manager of Clifton Springs Sanitarium Farm, in all of which positions he made good. He was Superintendent of Indiana State Experimental Poultry Station for two years before he came here, so he will be no experiment here.

Anyone contemplating going into poultry business on a large scale could put in a month under Mr. Kidd and learn many things that would be invaluable to them in their own business.

I intend to enlarge my buildings during the summer and raise double the chicks I have ever raised before, so that I will be in better shape to satisfy fall and winter trade with strictly high grade free range stock with a vitality that can't be beat.

Visitors Are Always Most Cordially Invited to Inspect My Poultry Farm at Any Time. Trolley Cars Stop at Our Cobblestone Station, "DeGraff," Right in Front of Our Farm.

ORIGINAL RUBBLE STONE PUBLIC FOUNTAIN WITH COBBLE STONE PILLARS.

As I am laying out on my farm what promises to be one of the most ideal suburban villages near Amsterdam, N. Y., I will show some of my original ideas in this line that I built with my own hands this past summer. The above fountain is on "the Main Highway Around the World" which goes through the Mohawk Valley, and it attracts much attention from tourists as nothing just like it was ever seen before. Over 25 tons of stones were used in its construction, some weighing nearly a ton, and it is so designed that the fountain playing at top supplies two levels of water, which affords the best of drinking water to man, beast and bird and being up-to-date it has an attachment to fill automobiles also. A cement door on hinges closes a large hole in center air tight, so that anyone can go down 4 feet below the ground level and attend to all water and electric wire connections from all directions. The pillars have original idea tops, which are filled with flowers the year round, and have large electric globe which burn all night, affording light at the intersection of this new street which is being laid out. The Electric station shown in picture is also of my construction throughout, and is unlike any other building on earth in many respects, and viewed by thousands of people each day that pass on the electric and N. Y. C. railroads, proving a very attractive advertisement, as well as a permanent, artistic station in front of my residence, where the citizens of the village of "DeGraff" can take the suburban cars to any city in the Mohawk Valley. I am in a position to design and construct any style of Rubble stone architecture, that anyone might need, and give you a design entirely of my originating, being both practical and ornamental and something that will last forever, if built on honor, as these samples are. I am now building a cobble stone fireplace in my residence which will be different and better than any one ever built and will heat 5 rooms without interfering in any way with the homelike effect of the old fashioned fireplace, but it will utilize the heat which generally goes up the chimney. Write me if you are interested in this subject for terms and ideas.

GIVE REDS A TRIAL AND BE CONVINCED.

"DeGraff's Best Reds" are bred to lay,
"DeGraff's Best Reds" are bred to weigh,
You give them but half a chance,
They are "Dead Sure" to make it pay.

Lady Graffmoore, Palace Winner

Champion Amsterdam

My strain of Reds has been greatly benefited by three grand males, that I have bred from very largely in the past, "Champion," Boston Cup winner in 1907; "Amsterdam," best known Red male that ever lived; "Champion Amsterdam," winner of first cock, shape and color club special ribbon at Madison Square Garden, 1911-12.

"Lady Graffmoore," winner of color special at Palace Show, was the mother of three males I will use this year.

So many people are being humbugged on trapnested hens and 250 egg hen that I will give you a few facts about heavy layers. To begin with I claim that there is no fowl that is so universally acknowledged to be the best layers, especially in winter, as the Reds. While breeding helps, still the type of the hen and the way she is fed from a chick to maturity determines her laying capacity, and the way she is fed regulates the number of eggs she will lay, as a perfectly healthy hen is a machine that will turn out about as she is managed. The Red type of long bodied, wide legged, wedge shaped hen is the strongest type of a layer, and their heavy growth of feathers makes them impervious to cold, and they continue business all winter irrespective of weather conditions. Customers report wonderful results in laying from all parts of the earth, even getting a report of 193 eggs per hen from a shipment to Alaska. Give your hens a tight house, with drop muslin curtain, which should be open all fair days, plenty of litter to scratch in, together with pure fresh water to drink, grit, charcoal and oyster shells, and if they don't lay on the following feeding rations then you have not got the right kind of fowls:

I am breeding eggs for hatching purposes only, and I sell all my fowls lay, the year round, so that I value one strong fertile egg more than a dozen forced ones. A chick hatched from my hardy, vigorous hens will mature into a more prolific layer than a chick from a hen with weakened vitality that has been forced to lay 200 eggs one year and won't lay half that the next year.

The fertility of eggs can be guaranteed at all seasons of the year as there is no breed of fowls so gallant and prolific as the Reds.

Photo of my $25,000 hen raising a promising bunch of young ones under most favorable conditions.

MY METHOD OF RAISING PRIZE-WINNING, HEAVY LAYING, STRONG VITALITY, RHODE ISLAND REDS.

Many people seem to consider the raising of chicks more a matter of luck than of management, but after many years of practical experience, I have decided while luck may have considerable to do with your success, as in all other undertakings, I find good management, looking after little details, keeping things sanitary as possible, and following nature's ways as near as possible, will produce the most high grade chicks of the quality I am after. There will never be any method of raising chicks to beat the old hen especially if she is of the right size and disposition. While I make use of the machines for early hatched chicks still after I am able to get setting hens I hatch all my best eggs under hens that are set on the ground, and after being kept quiet a day or so with the hen I give them free range to roam with the hen and scratch for their feed and learn to look after themselves as early in life as I possibly can, and you would be surprised to see the difference in health and vitality of chicks raised under such conditions as compared with incubator stock raised on bare ground in limited yards. I have decided it pays to have everything as near right in raising chicks as possible, as the number saved from the ravages of their many enemies will pay for the extra cost of protecting them.

There are many successful ways of feeding young chicks which can be adopted according to your local advantages, but I find that it is a good plan to always keep the chicks as warm and quiet as possible for at least three days and give them a good chance to get their digestive systems properly developed to do the required work.

The first feed can be crumbled stale bread crumbs, broken soda crackers, either dry or slightly moistened with boiled water or boiled milk. I advise feeding sour milk to young chicks right from the start as the acid in it is as good a preventative as we have for "White Diarrhoea"

Milk in any form grows fine chicks, but when cooked in some way it is better, especially if mixed with other feed, like corn meal or shredded wheat.

The regular ready mixed chick feed composed of all kinds of mixed grains is a good feed once a day for scratching, but you will find it pays to have a mixture of mixed ground feed composed of as many different grains as possible, and by all means have a good proportion of a good grade of meat scrap in it.

Green feed is very important at all ages, and it should be supplied in some form or other, as this is a very important part of free range chicks. After they get old enough to leave the hen, be careful that they have good comfortable quarters at night, and do not crowd too many in one place and thereby stunt the weaker ones.

As they get to the broiler age I kill all unpromising specimens and give the balance of the flock the best feed possible, such as hulled oats, kiln dried cracked corn, wheat or cracked wheat, and plenty of wheat bran and shredded wheat before them all the time, mixed with meat scraps. There is no healthier chick than the ones that roost in the trees, so I try to make my colony houses for maturing chicks about as near this style as I can, and give them protection from wind and rain.

Rose Comb or Single Comb Reds.

Be sure to mention which variety of Reds you want in each letter, Single Comb or Rose Comb, as it saves time and trouble.

MY CURES FOR POULTRY DISEASES.

Impure water is the cause of more disease than any other one cause. Occasional use of permanganate of potassium will be beneficial, as it is a great purifier, and boil all drinking dishes at least twice a month. Be very particular about drinking dishes for young chicks.

Roup, in its first stages, can be cured by use of kerosene in nostrils, mouth and eyes, and a two grain quinine pill at night will help some.

A teaspoonful of castor oil will help either diarrhoea or constipation.

Scaly legs can be cured by dipping in a mixture of equal parts of kerosene and lard, to which can be added a little carbolic acid. Dip two or three times, about a week apart.

Bumble foot can be cured, in slight cases, by iodine. Severe cases should be lanced, cleansed thoroughly, put on carbolic salve and bind up with cloth.

A pinch of common cooking soda will help indigestion and restore the red to a black comb.

I have found no better cure for chicken pox, canker in throat, or cuts from fighting, than to wash the effected parts in alum water. Take a cup of hot water and put in a teaspoonful of powdered burnt alum and after it is dissolved, wash all affected parts, as often as it seems to need attention.

Taylors Oil of Life sold by all druggists is a good all round remedy for many troubles that fowls are liable to have.

Mentholatum is good for colds in the early stages.

ACTIONS SPEAK TRUER THAN WORDS

You are judged, after all, in your journey through life,
By the gray matter under your hoods;
And the man who wins out in this strenuous life,
Is the man who delivers the goods.

HINTS ON FEEDING OF FOWLS.

The most important element of success in the poultry business lies in the man that carries the feed pail, as he can make or break the business by the way this part of the work is done.

While many advocate hopper feeding and others to make all fowls scratch for all they get, still I consider both ways all right if properly done, and neither will be a success if improperly done. Judgment has to be used in both cases and different fowls want different treatment. My ideal way of feeding is to have fowls rather hungry scratching for feed all day till toward evening, when they should have access to a hopper containing all kinds of feed, so as to go to roost with their crops full of the necessary, and while they do not always know what is best for them, still they do not make many mistakes in the long run. I think all fowls should go to roost with all they want to eat, as then is when the systems are built up and eggs are manufactured, and the difference between half grown chickens that go to roost only half fed and sleep in crowded quarters, unventilated, and chicks that have hopper feed at night and have comfortable quarters at night, properly ventilated, is remarkable, and proves it is very poor economy to try to save feed by feeding growing stock any less than what they will clean up. While wet mash will produce good rapid results for a short time, still I am in favor of a dry mash in hoppers if you have the hopper that works right. A mash in a hopper that is clogged up and they can't get a piece of it except when you knock it down to them is not a success.

The more grains you have mixed in the ground mash the better. I always like to use corn meal, oats, wheat bran, barley, peas and sometimes beans, together with meat scrap, or better still, meat meal, to which should be added some linseed meal at all times. The construction of the digestive organs of the common fowl is such that the mastication takes place after swallowing, and a crop is provided as receptacle for food in which it may be softened, thence it passes to a hard, muscular, hollow organ, where it is thoroughly triturated and reduced to small particles and rendered digestable.

These peculiarities in the construction require an increasing amount of hard substance to be taken in with the feed in order that the organs used for the mastication may be gradually developed so they will eventually completely perform their functions.

I know of no better grit for this purpose than a heap of sharp, gritty sand, and this, like pure air, is cheap, and these two items are more important than any others, still they are most often overlooked.

While fowls are naturally scavengers, still they do better when fed on good sound grains, and the eggs are of a very different texture and taste. The following grains are all good and you can select those that you can get to best advantage in your locality:

Clean, bright wheat, cracked or whole, is probably as good a feed as we have for either growing or full grown fowls. Although rather more high priced, still it produces results that make it cheaper feed in the end.

Shredded wheat can be bought in bags now and it makes a very fine feed when fed in the hoppers with ground feeds, and being nothing but whole wheat, I can't see how any food could be more healthy.

Oats is one of the best balanced feeds we have, however, the coarse, fibrous hull should be removed for young stock, and it pays to buy oats of the best quality so the heart will be heavier in proportion to the hull.

Rolled oats, or cracked oats, is one of the most important foods we can feed to growing stock, as it is a muscle builder and makes fine feathers.

Corn is the most common article of food for fowls, in fact many people in years past never fed their fowls anything but corn from hatch to death, and had comparatively good results, but such feeding would require free range to balance up such one-sided feeding.

Corn is rich in carbo-hydrates, hence is largely consumed in production of heat and energy, for this reason it is a very desirable feed in winter time; still it should be fed in connection with other grains.

Barley has much the same combination as oats, but is slightly richer in protein, and has the advantage of having only one-third as much fibre.

Oil cake or ground flax seed is also a very good food for growing fowls and laying hens, especially during moulting. It is one of the richest foods we have in bone and feather forming materials, but is too rich to be given alone and should not constitute more than 10 per cent. of the ration for fowls.

Peas are extremely rich in protein and cracked peas can be added to any mixture of grains to advantage.

The value of charcoal is very often overlooked and it should be fed to fowls of all ages as a preventive of disease. It should be kept before fowls at all times, as there is no danger of them eating too much, it has a great purifying effect in absorbing noxious gases and will correct many digestive disorders.

HINTS ON BREEDING REDS.

As the male is half the pen he should be about as near the Standard requirements as possible, but no matter how good he may be, if not a vigorous, healthy bird he is worthless, and should be discarded at once. While the old saying, "Like produces like," may be true to a certain extent, still it often produces the good or bad points of ancestors, several generations back, instead of first parent, which makes the breeding of Reds especially difficult, as they are made up of crosses.

I have made a practice for several years back of not using any bird in my best pens that I would not like to see reproduced, so that my breeding pens of today have nothing very bad to strike back to for several generations, which accounts for the wonderful results that can be obtained by buying one of my line bred males.

Avoid all extremes of color, as they seldom nick and are liable to produce mottled and badly off colored young stock. The most desirable male is the male that is about medium in shade of red, but very rich in undercolor and very brilliant and harmonious in all service color, and with a rich breast, as the breast is very important in deciding the even color of your pullets.

The female regulates the shape and size, but we must look to the male for vigor and color, but the nearer we keep to both sides of the union, the better average will be the flock of young stock.

Select the females that are strong in the points that the male is weak in, so that they will strengthen his weakness, and in fact split the difference, as it were, and produce chicks that are better in all respects than either of the ancestors.

If possible have the females match the male's breast and you will be assured of a flock of young stock that is not very far from right. Many overlook the other desirable points in mating for color only, but you should give due consideration to the oblong bodies, and the black markings of the tail and wings, especially the outer edge of the primary feather of wings.

While the Standard calls for an edge of black in the wings of females, still very few of the finest colored birds ever have it right, but when it does appear it is a great point in the exhibition bird, and should be aimed at, for a Red female tipped off with perfect black markings is a bird hard to equal in beauty in the show room.

Every breeder should have a copy of the Standard of Perfection issued by the American Poultry Association and then you can judge your own Reds as good as anyone else, but as many I receive letters from claim they have never seen a good Red, I have given a few of the principal points to be considered in selecting your show birds or breeders.

My color plate will give you a better idea of the ideal shade of Red than any word description, and while this is ideal, still many of the best show birds are not quite up to this ideal in color.

The prize winning male is generally the bird that has the most brilliant, harmonious surface color, if he is all right in other respects.

MY COCKEREL CONDITIONING HOUSE

As I sell so many high class breeding males to improve the flocks of others all over the World, I have to keep about 200 males of all ages in a separate house, which is designed just for this purpose. The lower floor takes care of the regular stock, while the next floor, has smaller pens for special birds, all of which have all care possible to promote perfect development, and kept in best of condition.

Twenty-five separate coops, about 4 by 5 feet each, are so arranged that the birds cannot possibly damage themselves in any way, and each birds has its own drinking dish, filled automatically, and electric alarms and electric lights make the house complete in all its equipment.

The score card is the worst farce of the poultry industry. I introduced this record card at Springfield, Ill., School of Judges. I use it at shows I judge at and send it with high priced birds that I sell. Different judges scored the cockerel listed here all the way from 88 to 94.

THIS IS THE GUY THAT PUT ED IN REDS

HEAD OF THE RHODE ISLAND RED INDUSTRY.

Being naturally modest and of a retiring disposition, it goes against my will to print this, but as business interests demand it and the children must have shoes and the fowls feed, I will have to overlook my personal feelings and let the good work go on.

DE GRAFF POULTRY FARM, AMSTERDAM, N. Y.
Leading Single Comb and Rose Comb Rhode Island Reds Specialist of America

MY EVIDENCE IS ALL IN. WHAT WILL YOUR VERDICT BE? FOR DeGRAFF OR AGAINST?

WHENEVER YOU SEE THE NAME "DeGRAFF" THINK OF RHODE ISLAND REDS, AND WHENEVER YOU HEAR RHODE ISLAND REDS, THINK OF "DeGRAFF," AS I HAVE TRIED TO MAKE THEM SYMBOLICAL

LIFT THE DeGRAFF BAN

The American Poultry Association has no moral, and probably not a legal right, to hold over the head of DeGraff a life sentence, and he should be quickly and speedily restored to membership.

If the entire membership of the American Poultry Association was to vote upon the question of reinstating DeGraff, we doubt not but that there would be an overwhelming majority in favor of it.

Poultry Success is carrying Mr. DeGraff's advertising, and in doing so we, as per the terms of our guarantee, are assuming every risk that our readers may take in patronizing him.

However, in accepting Mr. DeGraff's advertising again, we did so with the utmost confidence in him, not only as a breeder of Rhode Island Reds, but as a man of integrity and good intentions.

DeGraff is determined, and seemingly destined, to become one of the greatest factors in the country for the upbuilding of the poultry business in general, and particularly such breeds as he may now or hereafter become interested in.

Mr. DeGraff is possessed of rare ability as a writer on poultry subjects, and as an advertising man; in fact, is a genius in many respects, and the poultry industry needs just such men as these. It is just such men as these who have in times past revolutionized thought and action and thereby shaped the lives of men and nations.

Lift the ban. Even those who were active in the prosecution of DeGraff now recommend such action on the part of the American Poultry Association.—Henry Trafford, Editor Poultry Success, Springfield, Ohio.

SEND FOR FALL SUPPLEMENT PRINTED OCTOBER 1, 1913

Once a year I issue 5,000 of these books in which I insert a business supplement. I have a mating list for spring business and fall supplement for fall and winter trade. Please write me and explain what you need, before buying elsewhere, as there is no breeder on earth in better shape to serve you.

MY EDUCATIONAL CAMPAIGN FOR AMATEURS.

I have tried through all these 24 pages of hen talk to give you as much educational, original, practical information as I possibly could. I could have given 24 pages of testimonials from satisfied customers, but I give only a few clippings from prominent people who are in a position to competently speak of my business proposition. If you are in sympathy with my style of writing I would advise you to subscribe for Rhode Island Red Journal, Waverly, Iowa, (25 cents a year), as I will have an article in this excellent Red paper each month. I will also have a well-illustrated article in POULTRY, published by Poultry Publishing Company, Peotone, Ill., each month, in which I will give all Red breeders the latest and most interesting facts pertaining to Reds as I see them at largest shows and otherwise. Send me 50 cents for a year's subscription ($1 regular price). Let us pull together for true reform, as it is high time some of the papers were published in the interest of the reader, as well as the advertisers, as the unsophisticated amateurs have been sadly abused.

THE HUSTLING CITY OF AMSTERDAM, N. Y.

Although Amsterdam is not one of the largest cities in the state, still it is one of the most enterprising for its size. It is situated in the beautiful Mohawk Valley about half way between Albany and Utica, on the lines of the N. Y. C. and West Shore railroads. The streets are well paved, and the surrounding country has macadamized roads in all directions, while the electric car service is as good as any city in the country, especially the interurban limited cars connecting with Gloversville, Johnstown, Schenectady, Albany and Saratoga Springs, are the finest in the world.

It is strictly a manufacturing city, and employment is given to all classes of workmen, and good wages paid.

A very good idea of the place is given by the following facts:

Amsterdam Has and Is:

Population 35,000.

The largest Pearl Button Factory in the World.

Is the First City in the World in the Manufacture of Brooms.

Is the Second City in the United States in the Manufacture of Carpets and Rugs.

Is the Third in the Manufacture of Knitted Goods.

Is the Ninth City in the State in Value of her Products.

Has eighty-nine manufacturing establishments, whose annual output amounts to $15,276,000. Still the only advertising matter ever sent out of the city that was high grade enough to sell, at 25c. a copy, is the catalogue of the DeGraff Poultry Farm, whose mail is one of the largest received by any one firm in the city.

I Stand Ready to Take Your Orders for the Best Rhode Island Reds on Earth, and Guarantee Safe Delivery to Any Part of the World

Leading Single Comb and Rose Comb Rhode Island Reds Specialist of America

YOUR MONEY'S WORTH OR YOUR MONEY BACK.

I AM DOING BUSINESS "ON THE LEVEL" WITH THE WHOLE WORLD.

I want to tell everyone in the world interested in better poultry that the DeGraff Poultry Farm of Amsterdam, N. Y., E. U. A., is in better position today to fill your orders for anything in the line of Rhode Island Red fowls or eggs than ever before in the history of this farm.

For over 25 years I have made it my life work to build up an establishment that would be in a position to sell the highest grade of poultry that could be shipped to all parts of the earth for the improvement of others, and I now feel that I have reached that stage. The Reds have demonstrated through their many practical money making qualities, that they are the Greatest Utility Fowl on Earth, and enjoy the greatest popularity today ever given any breed in America while their popularity is fast spreading to all countries of the globe. All honest fanciers will admit that I have done more for the advancement of the Reds than any two breeders and this together with 10 years of extensive showing and advertising, has given me a reputation known all over the world as "The Leading Red Specialist of America" and I am now able to sell all my stock and eggs without showing or extensive advertising, thereby cutting out the two greatest leaks in the poultry industry. I have always tried to deal liberally and honestly with all customers, as I find this the cheapest and best advertising that I can have.

I have exported Reds to nearly every civilized country of the globe and it is astonishing how they adapt themselves to both cold and hot countries equally well, and I have failed to find a case where they have not made good.

The possibilities of this business are unlimited, and when you think of the number of fanciers all over the world that are sending to me each year for new blood to improve their breeders, both as to show and laying qualities, you realize the importance of doing everything in my power to improve my own flock. Many prominent breeders look to me each year to supply their winners, and the fact that I have for years furnished winners of first prize, and shape and color club ribbons at Madison Square Garden proves conclusively that I have all I claim and more.

Make all remittances payable to

EDWARD T. DE GRAFF,
Amsterdam, N. Y.

Bell 'Phone 94-J.   Cable Code—Redgraff.

PARCEL POST DELIVERY OF EGGS

The Parcel Post is in the experimental stage, and what results it will deliver cannot be known till tried. I will agree to deliver by this system all eggs ordered at $10 per 15 eggs or more, and all eggs listed less than $10 per 15 at 25c. a setting extra.

I will use the most practical method possible to secure and give it a fair trial and hope for better results than by express, especially in small orders.

THE A. P. A. REFORM GOAT

Four years ago I was the victim of an unprecedented action taken by the A. P. A., commonly called the "DeGraff Incident," which has done me irreparable injury the whole world over, although the foundation on which it was based was as insignificant as anything could be imagined, and worse offences have occurred and do occur today at every show that is held, and go unmentioned. Although my application for reinstatement has been tabled, I am informed that in order to be a member of the A. P. A. I will have to remit $10 and get signers for membership blank. I have positively refused to pay $10 again. All I got before was $10,000 worth of damages for last contribution, and it is immaterial to me if I ever join the A. P. A. again or not. If they can consistently give a member a life sentence for what every other honest member admits he does, then the longer this inconsistency goes on the more the injustice is realized by all well posted fanciers.

Just to carry on this unprecedented farce, I will suggest that every satisfied customer of mine, or friend of fair play, remit S. T. Campbell, Secretary of A. P. A., Mansfield, Ohio, a one cent postage stamp towards my required $10, accompanying the letter with your experience on this subject, and if the $10 is ever secured this way I believe it will prove that I am giving my customers their money's worth, and partially vindicate my contention.

It is not a matter of $10 to me, but a matter of principle, and I know from letters received that there are a thousand and more that will willingly do this to prove that they feel I have been a victim of misplaced justice. This is a year of "Down With the Bosses; Let the People Rule," and there is a chance for reform in poultry circles as well as politics and business. Testimonials enclosed prove that I am as reliable as any breeder doing business today, so I will give this statement as my last word on this subject, let future results be what they may.

SPRING SUPPLEMENT TO DeGRAFF'S BOOK ON REDS

I consider the accompanying testimonial from I. W. Bean the most convincing argument that I can give to any prospective customer that I have what I claim, and that I am in a position to sell you what you want to buy better than any other breeder on earth. I don't believe there is any Red fancier that is as universally acknowledged to be the accepted undisputed authority of Reds as this man Bean, while his honesty and good judgment, in giving me this unconditional write-up, would not be questioned by anyone that knows him.

While the world is full of "knockers," it is a pleasure to meet a man that has a good word for others, when they know it is deserved.

Among other exceptionally good Reds sold by Mr. Bean may be mentioned his famous "Sensation" and Tompkins $2000 male, while hundreds of breeders of both breeds acknowledge their flocks have been benefited by Bean blood. The original light hackle Rhode Island Red has been superceded by the "Bean Red," which has the dark hackle, with

all sections harmonizing.

PERSONAL VISIT OF I. W. BEAN, OF BRAINTREE, MASS., JAN. 26, 1913.

To Whom It May Concern:

While I have known for a long time personally and sold many birds to E. T. DeGraff, of Amsterdam, N. Y., still I have never been able to visit his farm till today, and I feel well repaid for coming, as there are many features about this farm that I believe can not be seen elsewhere. While at Boston Show I accepted his invitation to inspect and assist in mating some of his best breeding pens for 1913, and after handling some of his best birds, and taking into consideration the large number of Reds of both varieties, none of which I could say were not fit to breed from, I feel safe in saying I could not dispute his claim of having the "largest and best flock of Reds on Earth."

I would advise anyone contemplating going into the fancy poultry business to visit this farm, as here can be seen many original styles of houses and labor saving devices which cannot be found elsewhere, that will well repay you for the trip, and incidentally see what a large flock of high grade Reds really look like in their natural element.

Like many others my curiosity had been aroused by the many good and bad things I have heard about this man DeGraff, and for the information of others who can not personally inspect this farm, I will state the following facts, as I saw them:

I found all fowls housed in open front houses with muslin frames attached, and I failed to see a sick bird in the flock, which numbered over 1000 birds and judging by brilliancy of plumage, and general redness of flock, I believe the fertility of eggs sold from these birds should be very satisfactory, as no forcing for egg production during winter is practiced, so that a full crop of strong germed eggs will be laid in spring for hatching. I personally handled pens of 10 to 12 females each, that would average in quality with some of the prize winners at Boston and New York Shows, while the pedigree specimens are about as strong breeding quality as I have ever handled.

After inspecting what will be the 50 breeding pens as listed in circular, comprising S. C. and R. C. Reds, all headed by first class males selected to blend to best advantage with females, in each pen, I was shown his cockerel house in which I saw over 200 males of various ages to be sold and shipped to all parts of the world.

Anyone who doubts the facts that Reds can be bred in large numbers and have what can really be called red, should inspect this bunch of dark necked harmonious males, as the three colored birds are noticeable by their absence. The flock of hens I saw here will average darker than any large flock I ever saw, and should make very valuable breeders judging from their type and appearance.

There are many smaller breeders who have fine specimens in their flocks all over the country, but the one point that impressed me here, was that DeGraff not only had his special pens to produce his stock for next year (and he will use many different pens himself) but that he has so many good birds that he can spare eggs or stock from his very best, to those who are willing to pay what they are worth, **and not cripple his own flock.**

DeGraff claims he has been working for over 12 years to get in the position to be able to do a legitimate fancy poultry business on a large scale, and endeavor to give every customer a square deal (as near as the shortcomings of nature will allow) and I believe when the fact of his large flock is taken into consideration with his untold educational stunts, he has pulled of for the benefit of the breed, that his advertised claim of "The Leading Red Specialist of America" is not an idle boast.

While I have never been an extensive advertiser I was much interested to see the practical results that can be accomplished by extensive advertising, which he has done in a style entirely original with himself. I saw a frame containing letters from apparently every civilized country of the Globe addressed to this farm, showing his campaign of education to the good points of the Reds, has reached every part of the Earth. Of the 90 letters received the day before, I noticed about 25% were postal inquiries, 25% regular inquiries about orders, and 50% contained quarters in payment of his color plate catalogues, showing people are willing to pay for a high grade article, even in the advertising line, when properly put up for them.

I can readily believe, that it is true, that no breeder in the last six years has ever kicked on his investment of 25c for this Book on Reds after he had seen and read its contents. From what I saw of the prospective "DeGraff's 1913 Book on Reds" it will be the finest thing of its kind ever printed, and contain information and pictures that no breeder can afford to be without, and more pictures will be framed and preserved than will ever be thrown away.

There are few farms so naturally adapted to the poultry business as this one, and when he has his 500 foot winter house completed, which is now partly done, and more under construction today, together with the 600 apple tree orchard colonized for summer, all of which has running water piped to all parts, there will be little lacking to make this an ideal poultry establishment for carrying out the unlimited possibilities that this true fancier of undownable enthusiasm, has planned for the years to come.

I also saw and heard about the other branch farms that are raising DeGraff's Reds under his supervision, so that I believe it will be possible to fill any and all orders that may result from his extensive campaign of advertising planned this spring, with first class quality eggs even to 1000 a day if need be at certain times.

I trust this little testimonial to DeGraff, his Reds, his farm and his business methods may assist in some small degree in restoring a good friend, an enthusiastic and capable fancier and business man to that position where he will again have the confidence which he merits of his prospective customers.

He has been dealt with all too harshly in the few years just passed, but the tide has turned and now let those of us who believe in fair play and "a square deal for every man" be as ready to commend as so many unthinking ones have been to decry.

There is plenty of room for us all and DeGraff is surely coming into his own.

Yours very truly,

I. W. BEAN,
Braintree, Mass.

I DON'T KNOW HOW I COULD GIVE YOU A STRONGER TESTIMONIAL THAN THIS FROM OUR POSTMASTER.

Amsterdam, N. Y., Aug. 7, 1912.

The American Poultry Association,
S. T. Campbell, Sec'y:

Gentlemen:—

For the purpose of giving some information in regard to Edward T. DeGraff, with whom you recently had some misunderstanding, I will give the following facts:

I have known Mr. DeGraff and his father all my life, and they are well known citizens and one of the oldest families in the Valley. He has served over 20 years in the banking business. He has been Treasurer of Montgomery County Historical Society for over 6 years, and been Treasurer of several social clubs, which proves we can trust him.

He collected and expended the wheelmen's money for 8 years during which time he constructed over 15 miles of first class cycle paths without charging anything for his services.

He has received mail from all parts of the earth through this office for 10 years and to my knowledge there has been but two trivial complaints brought to the attention of the Post Office Department, both of which he attended to promptly.

Respectfully yours,

THOS. LIDDLE,
Postmaster of Amsterdam, N. Y.

INSTRUCTIONS FOR ORDERING EGGS.

Terms Strictly Cash in Advance—"Your Money's Worth or Your Money Back"—Write Your Name and Address Plainly.

Half Price on all eggs from listed pens after June 1, 1913.

Mention breed wanted in all orders and in each letter. Don't expect me to remember all back correspondence.

In ordering eggs I would advise giving at least 10 days' notice if you want any particular pen, and I will try to ship on date specified, but I advise you to leave the selection of eggs to me as much as possible, as I always try to give as good value as possible and often give eggs from better pens if I have any surplus. I recommend sending orders in registered letter, as this method prevents any misunderstanding or lost letters, as you get your receipt promptly and have proof that I have your order.

All eggs shipped from this farm are guaranteed to be fertilized by the union of two specimens as near perfect as possible to secure, carefully mated to produce the best results, by off-setting the weak points of one side by specimens strong in those weak points.

I am really proud of my flock of hens and the pullets that I am using are early hatched and well matured, so their eggs should hatch just as well as the hens, and as I always give my breeders a rest during the winter season, and house in healthy, fresh-air houses with free range in orchard, I can safely guarantee 80 per cent. fertility in all my eggs.

After many years' experience in packing eggs for shipment I have learned many things which help deliver the eggs to my customers in perfect hatchable condition, and if the package is not completely smashed by reckless expressmen, you will get your eggs all right. I have discovered a method of attaching a pneumatic cushion to the bottom of the package which relieves all jar from vibration of the cars, and I will use this on all orders for $10 worth or more of eggs. While the attachment is expensive and makes extra work, still I am willing to stand for this expense if I can give my customers better satisfaction.

I have shipped 540 Red eggs to California in one shipment and only 9 eggs were broken. I have shipped 150 eggs to Honolulu, Hawaii Islands, and had a 50 per cent. hatch. I have shipped 15 eggs to Scotland and had 12 healthy chicks hatch. I have shipped 30 eggs to Alaska and had 24 eggs hatch and 23 live, which proves conclusively I know how to pack and how to produce fertile, strong germed eggs at all seasons of the year.

As my laying force is so large and many pens are of equal merit I can fill orders of any size and furnish strictly fertile eggs. My pens are mated for hatching purposes all the year round and I take particular care that no eggs are chilled in houses or transit. There is such a demand for my eggs for hatching that I am selling nearly every egg laid, and most of the time I have orders booked waiting for the eggs to be laid from some special pens of extra quality. Everyone wants the best possible to secure. After June 1 and up to February 15 of the following year I always reduce the price of all eggs from listed pens one-half, which gives anyone an opportunity to get my best eggs at very reasonable prices, and many of the finest colored show birds are hatched late in the season.

I reserve the right to change males in pens, if occasion demands it, by placing other males of equal merit.

INSTRUCTIONS FOR ORDERING STOCK.

I will not give a table of prices on my stock, as it is not fair to lump all specimens under one head, as each specimen varies in value according to its individual merits, and the seasons of the year has much to do with it, as at certain seasons male birds are worth twice what they are at others.

It is better to buy one extra good pair than more inferior birds. I can mate you a breeding pen of any age of either breed, as you prefer, and give you a start that will produce you some very fine young stock next year.

After July 1 I will dispose of most of my 1913 breeding stock to make room for young stock, which will be an opportunity for all to get some really fine Reds at right prices.

I Stand Ready to Obey Your Orders Promptly

While I make a practice of killing all unpromising birds at broiler age, still in all large flocks there are birds of all qualities, and I can select either males or females worth the price you care to pay, and give you good value and the benefit of my extended experience. All birds shipped are banded with sealed leg bands, with number recorded on my books, and bearing the name of this farm, and I do not make a practice of shipping anything I would be ashamed to have others know they came from Amsterdam, N. Y.

As I have a very large stock to select from at most seasons of the year, I consider the fairest way to come to an understanding is for you to specify about what you want to buy, and I will give you my best prices on same at that particular time.

Many times a certain male that would not be of particular value to one person would be very desirable for some other order that required a bird especially strong in some certain points, so it is better to correspond than to set any iron-clad prices.

I never keep over a cock bird that is not worth $10 or more, according to what condition he is in or how old he is. While $25 would buy a cock bird fit to show, still I get as high as $50 each for my very best yearling males that are in first-class shape. In the fall I always have a lot of good breeding hens to sell cheap as compared to prices I would ask in the spring, and these birds, mated to young cockerels, make the very best start in the breed, as hens mated to cockerels make the best breeders, and both are cheaper at this particular time than any other.

I can sell you good hens at any price from $2 to $5 each in the fall and early winter.

I can make much better prices on stock before the time I have to house them, as it saves considerable work and room required to properly keep them.

I receive many letters asking me my honest opinion of the relative merits of the two different varieties of Reds, and my advice is generally for them to select the one that strikes their personal preference best, as there is supposed to be but little difference between them outside of shape of comb.

I have a regular cockerel house filled with individual coops in which I am able to condition each bird before shipping and then they are coop trained and act better when shown at any poultry show, which often helps decide the winner in a close decision. During the fall I can sell first-class cockerels for $5 each, while $10 to $25 will buy birds that I can guarantee to be sure winners at most local shows. I also have cheaper males at all prices in the fall before housing time, but when housing time comes I either sell or kill all birds that are not worth at least $5 each, so you can see it is to your advantage to order early in the fall.

In the fall I can sell you good pullets at $5 each, while $10 to $25 gets my best birds, fit to exhibit at any local show.

During the winter I have all qualities at all prices, until I mate my pens after the first of the year, when I am generally pretty well sold out.

I can give especially good value in mated breeding pens, as by selecting a male to breed with females selected to fill an order, I can guarantee to give better results than if I could only see one-half of the mating and judge from description about the other.

As all my stock is farm raised, you can depend on getting stock entirely free from all disease, and birds of a quality that will do your flock good if you have any, and if not I will give you a start which will guarantee you success in the future.

While I consider the average score card of little value, and when issued by some inexperienced judge of Reds, little more than a farce, still I will, if requested with order, furnish my record card with each bird costing over $5, and fill in the blank space with comments on what points I am cutting the bird, and why. I find there are many breeders that really do not know what a Red should be, and I am doing what I can to educate them, by furnishing color plates, and now I will give them the benefit of my experience by furnishing them Record Cards of merit.

Typical Pen of my Single Comb Red Matings

All $1 and $2 Eggs Guaranteed Fertile, and delivered free by Exp. or Parcel Post

PEN 1—MY FAVORITE PEN—BETTER THAN BEST

Eggs $2 Each in Any Number Desired

In every large flock of fowls there are certain specimens that stand out like the moon on a starry night, and it is these specimens that every true fancier uses to breed his next year's chicks, and will not sell at any price. I have selected this pen of hens and pullets as not only my very choicest specimens, most adapted to breed with what I consider the strongest breeding male I now own. While I will set every egg I can from this pen, still I will sell a limited number to anyone that will pay what they are worth, and will guarantee every egg to be fertile.

PEN 2—MY CHAMPION AMSTERDAM PEN

Eggs $1 Each in Any Number Desired

The cock that won First Prize, including shape and color club specials at Madison Square Garden 1911-1912 show heads this pen, mated to 10 pullets of his blood lines, that can not help but produce winners for 1913 shows. I exhibited this cockerel in a display cage at Albany and Schenectady, and I never saw a bird in my life that was so universally liked. Later when I sold him to Charles Whitney of Albany, and he won the Championship of the Madison Square Garden show, I was not surprised, and when Judge Bryant pronounced him one of the finest Red Males he had ever seen I was pleased to know our ideas agreed so well. Realizing his breeding value I bought him back to head one of my best pens for this year, and I consider this a guaranteed winner breeding pen and will set every egg I do not sell at $1 each.

PEN 3—DE GRAFF IDEAL TYPE PEN

Eggs $1 Each in Any Number Desired

Many breeders in their craze to get color have overlooked type, and now have birds that severely lack in vitality and size and are of undesirable type. I have selected 10 of my best hens of several years back, and mated them to the most ideal shaped cockerel I own, so there can be no possible doubt but what anyone in need of a good male of good type to cross with their stock next year can not make a better investment than in some of these eggs from this pen. While this pen is unusually strong in all points I consider it almost ideal in type, and it is fast coming to be realized that type should be considered half of every prize winning Red.

EGGS FROM INDIVIDUAL PENS OR ASSORTED FROM ALL THESE PENS

Guaranteed Fertile Eggs
$10 per 15; $18 per 30; $25 per 50
FROM PENS 4, 5, 6, 7, 8, 9.

PEN 4—MY IDEAL COLORED PEN

Every breeder has some weakness in his flock that needs strengthening, and I had the thought in mind when mating this pen to mate up a pen of exceptional strength of color in all sections that anyone in need of more color in their flock could buy these eggs and be sure of getting extra value.

While they are especially strong in color, still they are right for size and shape, and no breeder can go wrong in selecting them.

PEN 5—MY LARGE TYPE PEN

I find many breeders are looking for something to improve the size of their flock, so I mated this pen especially to help these fanciers out, as all the hens that compose it are over weight, and still very strong and active.

The cockerel heading this pen is one of my very largest and has the developments of a cock bird, although still young.

PEN 6—MY IDEAL BLACK MARKING PEN

Many of the principal prize hens and pullets of the country lacked in black in the wings and tail. If the breeder showed a bird equally as good in surface and still had wing would have won out, so it pays to watch your females' wings all you can, as it is a strong point with some judges. Try this pen for wings.

PEN 7—MY VITALITY PEN

No one denies that old hens lay the strongest germed eggs, and especially when mated with a young cockerel of exceptional vigor, do we get those large fluffy chicks that nothing can stop them from living and growing like weeds through all kinds of neglect which might kill off many weaker chicks.

This pen has type and color and is especially selected to give you the vitality we are all in need of.

PEN 8—COCK OF THE WALK PEN

This three-year-old bird is one of the most vigorous birds of his age that I own, and he will fight for the rights of his family at the drop of the hat. He has my ideal type with good low tail well spread, making a grand, good bird in every way, and has proved himself a good breeder in years past, and should do even better, as he is mated this year.

PEN 9—MY HARMONY PEN

This pen of pullets fully developed, is headed by a cock that has always attracted the attention of all visitors, as they have never seen such harmony in any bird in their life, as he is simply one shade of red all over, and has bred the same in his young stock. If your stock is in need of improvement in surface color try some of these eggs and you will never regret it.

Guaranteed Fertile Eggs
$8 per 15; $15 per 30; $20 per 50
FROM PENS 10, 11, 12, 13, 14, 15.

PEN 10—MY PEN FOR GOOD HEAD POINTS

Some breeders have great difficulty in getting birds with good combs and eyes, so that I have selected this pen of birds that anyone could give them a trial, and feel sure of getting what they needed, as every bird in this pen, while strong in other respects, has as perfect heads as you often see on any birds.

There is no point appreciated as much in the new beginner as a good comb, and every one likes to see them as near right as possible.

PEN 11—MY UNDER COLOR PEN

There is no one point that signifies a strong breeder as much as the under color of the feather, and especially the red quills. In selecting this pen I used only birds of exceptional rich under color with very best of color in quills, so that if your flock lacks in strength of under color, try a few of these eggs for a cross male for next year, and get some show females as well.

PEN 12—MY LADY GRAFFMOORE PEN

The females in this pen were selected to mate with this most promising cockerel sired by Lady Graffmoore No. 2. This bird has many good points that can not help but make him an exceptionally good breeder, but I am banking on the breeding back of this family to produce some of my very best breeders, as they have nothing bad to strike back to for several generations.

PEN 13

This pen is composed of 10 as good typed hens as anyone could wish to see, and although lacking in surface color, still they have exceptional under color for hens, and when mated to this son of Palace winner, you are sure of as rich under colored chicks as could be bought at any price.

PEN 14

This is another pen of my very best hens selected to mate with a cockerel from Lady Graffmoore 2nd, and makes a back fire cross that can't help but produce a large percentage of good young stock. A cockerel from this pen would make a good breeder next year, even if he himself did not have many show points, as the breeding back of him could not help but show up later in breeding.

PEN 15

The cockerel in this pen is a son of Lady Graffmoore, 1st, and is therefore one generation nearer my own stock and makes a good cross with the females I have used, which are partly the same blood. This pen should make a grand pullet mating pen, and if you get one pullet from a setting you will get great returns for your investment for a breeder.

HALF PRICE ON ALL EGGS FROM LISTED PENS AFTER JUNE 1 TO FEBRUARY 15, 1914

In listing my matings I have not tried to describe each male that heads the pen, as it is useless repetition, as I can safely say right here that there is not a male in any of my pens but what is a first-class specimen, in both type and color, and naturally adapted to breed to advantage with the females he is with.

There is not a bird used that has any white in under color or any section, no white earlobes, or anything but red eyes can be found, while the dark, brilliant harmony color predominates the whole flock, and very little smut is used, and that only when it will be to advantage to use it. I do not specify prize winning birds, as I do not show, but as Mr. Bean says I have many birds that are better than many prize winners of the past year's largest shows, and all the better for not being shown, as they are in perfect health, for breeding, and not getting over effects of showing like many birds are that are praised so highly in some circulars. It is always a pleasure to show my pens to visitors, and everyone is invited to inspect them at any time.

Surface and under color of hackles

Surface and under color of wing bow and back

IDEAL COLORS OF R. C. RHODE ISLAND RED MALE

DeGRAFF POULTRY FARM, AMSTERDAM, N. Y.

Typical Pen of my Rose Comb Red Matings

All $1 and $2 Eggs Guaranteed Fertile, and delivered free by Exp. or Parcel Post

Guaranteed Fertile Eggs
$10 per 15; $18 per 30; $25 per 50
FROM PENS D, E, F, G.

Guaranteed Fertile Eggs
$8 per 15; $15 per 30 $20 per 50
FROM PENS H, I, J, K.

R. C. PEN A—BEAN COLOR PEN

Eggs $2 Each in Any Number Desired.

This pen is composed of some of my very best hens and pullets that are individually strong, and still just the right type and color to mate right with a yearling Rose Comb Cock that I bought of Bean just before his visit for this special pen. The color of this cock is about as strong as I ever saw and has no weakness in any place, and should breed as high grade chicks as money could buy, and the breeding back of this pen will improve any flock.

PEN B.—MY FAVORITE ROSE COMB PEN

Eggs $1 Each in Any Number Desired.

This pen is composed of ten about as fine Rose Comb pullets as I ever saw together, as they all have type and color to spare, so that I feel sure there will be great results from this pen, and don't care whether I sell any eggs from them or not, as I can see some next year winners coming from this stock already.

The cock bird heading this pen was also bought of Mr. Bean direct and was mated according to his directions, as he had proved a guaranteed breeder on just this same style of pullets last year.

PEN C.—MY GUARANTEED WINNER PEN

Eggs $1 Each in Any Number Desired.

This pen is composed of 10 yearling hens that produced some of the highest grade young stock of any I ever handled last year. I have them mated with my very strongest cockerels raised last season, and I feel sure they will do better this year than last. Anyone wanting to get a well bred cockerel to head their next year's pen could not do better than to try some of these eggs.

PEN D.—N. Y. STATE FAIR PEN

The cockerel heading this pen was the head of the First Prize pen at N. Y. State Fair last fall, showing he was a very promising young bird and he has developed into a grand bird when matured.

I don't consider him any better for winning a prize as I have better birds that were never shown, but some people like to buy of prize winners, so I list this pen this way. The pullets in this pen are early hatched fully matured birds like hens.

PEN E.—GRAND CENTRAL PALACE PEN

The cockerel heading this pen was head of third prize pen at Grand Central Palace in strong competition, and he is an especially strong bird for type and color, and backed up by best of breeding. The pullets I have in this pen are not related to they have bought of me in years gone by, chance to get a cross to go with any stock male and this would give my old customers a while anyone can make good use of them either for showing or breeding, as I consider this a very strong combination.

PEN F.—MY HARMONY PEN OF ROSES

There is no show point that helps out a winner so much as that velvet finish that is sometimes seen in birds when in pink of condition, and just the right age to show to advantage, but there is a great difference how long different birds will hold this color. This pen of pullets have held their color so far as good as any show birds I ever saw, and I consider they will breed it in their young and make a very desirable choice for anyone in doubt as to what pen to select eggs from.

PEN G.—MY BLACK MARKINGS PEN

This pen of pullets mated to a cockerel of exceptional color throughout, were selected on account of their perfect wing and tail markings, and anyone in need of more black in their flock should try this pen.

PEN H.—MY UNDER COLOR PEN

This pen of pullets were selected on account of their perfect under color, and while strong otherwise they are exceptionally good in this respect.

They are mated to one of my best cock-color, so that anyone in need of better uncterels selected for his strength of under der color should try one setting at least from this pen.

PEN I.—MY PERFECT TAIL PEN

There are many families of Rose Comb Reds that carry their tails too high, and some are badly pinched making very undesirable breeders while others have too much red where black should be. I selected this pen so that every bird in it should have tails as near perfect as any could expect to see, and the chicks from it can not help but be strong in this respect and be strong breeders.

PEN J.—MY COUNTRY JAY PEN.

This pen is composed of some of my oldest hens that have done good service in years gone by and although they may look like Jays, still like some country jays are the real genuine article and will breed more good birds than the fine young undeveloped pullets that have never been tried. These birds were good color as pullets but have faded from heavy laying.

PEN K.—MY DARK NECK PEN

Many claim the dark necked females are the best layers, but whether this is true or not, I do know that they breed some grand cockerels when mated right, and as I have this pen mated I am sure of some grand even colored cockerels and dark necked pullets that will help anyone for their shows or next year's breeders.

80 PER CENT FERTILITY GUARANTEED ON ALL EGGS ALL THE YEAR ROUND

SPECIAL SALE OF 1913 BREEDER IN AUGUST AND SEPTEMBER AT BARGAIN PRICES.

The reduced prices on my eggs after June 1 always carries my egg trade well along till middle of July, and after that I will sell a large share of my 1913 breeders at bargain prices; and there is no better investment in starting with Reds than a pen of yearling hens mated to a grand cockerel. I will have some grand cocks to spare also that will make both show birds and exceptional breeders.

Guaranteed winners for fall fairs; a special feature for fall trade.

MY AMATEUR HOPPER OUTFIT FOR $10.

The Sexton Dry Feed Hopper is as far ahead of the Boston Hopper as you can imagine. It is a complete success, and every breeder should use one. I find it also makes a very convenient egg carrier, so I have decided to make up a combination bargain for your benefit, and you get all these invaluable articles for one express charge.

30 of my best eggs of either breed (my own selection)	$10 00
1 Galvanized Dry Feed Hopper	1 00
1 Quart or Pint Chick Drinking Bottle	40
1 Box DeGraff's Condition Pills	25
DeGraff's Chick Punch	25
DeGraff's Book on Reds	25
Carefully packed in one package for $10.00	$12 15

STANDARD PENS OF BOTH BREEDS.

As it is impossible to give photographs of all breeding pens I have shown the above as my ideal type, from actual photographs. All pens are composed of 10 to 12 females to one cockerel, properly housed and kept separate, and as there are nothing but Reds on the farm you are assured of pure bred stock, of a quality in proportion to price paid. Besides the pens listed below, I have several other pens of both breeds which I call my Standard Pens, composed of the balance of my flock, and as many pens are headed by prize winners not mentioned here, and nearly all females are bred from equally as good stock as the others, you are assured of as good value from these eggs in proportion to price paid as from the following higher priced pens, as they are practically the cream of the country, and command the very highest price.

STANDARD PENS OF BOTH BREEDS.
Strictly High Grade Quality
$5 per 15; $8 per 30; $20 per 100.

HIGH CLASS INCUBATOR EGGS FOR BOTH BREEDS.
$3 per 15; $5 per 30; $15 per 100.

REGULAR INCUBATOR EGGS.
$10 per 100 at Special Seasons. Write.
You can divide order as you see fit.

A RED-HOT CHAT ON SHOWING RHODE ISLAND REDS

As there is no subject more interesting or important to the true fancier than his prospective show record at some local show, I will give the amateur a few confidential pointers which I have learned through years of experience in breeding and exhibiting Reds.

The unsophisticated amateurs who are learning the inside working of the poultry industry and generally believe all they read in the poultry press, always arouse my most sincere sympathy, for if they bite on many of the "get rich quick" schemes promoted therein they would soon be entirely broke and the honest element of the business has to suffer in due proportion.

If the following hints will help to put them wise to what they will inevitably learn by experience, I will feel well repaid for my trouble. I find, in the long run, that it pays all fanciers to pull together rather than to knock your competitor and his stock, which is too often the case in many localities I have visited.

As I have practically graduated from the showing and advertising stage of the poultry business with a reputation all over the World that disposes of my stock, I feel I should give the new beginners all the information I can, that will help them with my favorite breed, that they may make them the more profitable. While I do not say it egotistically, still I know it cannot be denied that I have spent more time, thought and money in studying what constitutes the ideal Rhode Island Red fowl than any other breeder on earth.

My color plates published for years were not accidents, but the result of the best brains in their respective lines carrying out my ideas for the benefit of every breeder who ever owned a Red.

The unprecedented popularity of the breed, can be greatly damaged by the many uncalled for displays of jealousy which have taken place in the Red Clubs, and I sincerely hope that every breeder will henceforth join the Club, and help elect officers who will do all in their power, to hold the breed before the public, and educate them to their many good practical points.

Red Hen Tales for 1912 is a move in the right direction but could be improved for 1913 by having more illustrations and less repetition in articles, and by all means cut out those articles teeming with personal advertisement and blarney.

I firmly believe in following nature for our ideals, in preference to the imaginary drawings and I believe in due time that the Standard will be illustrated with pictures from life, much to the enlightenment of the amateur.

Although I had to handle practically thousands of birds to make my selection of what I felt was my ideals, still I used actual specimens to make my direct photography pictures, and they are pronounced to be the only colored pictures that are at all true to life, and give the amateur the true ideal color better than any words can describe.

The poultry papers are educating their readers up to an impossible standard by showing drawings or retouched photographs, by the leading high priced artists, who manipulate the original photos beyond recognition.

I will show some actual photographs of really good birds of different types, with comments on each so we may better understand each other and I believe if other breeders would advance their ideals along this same line and show actual photographs of their winners at this winter's shows, it would help the breed immensely.

These companion pictures show the actual photograph and the retouched one after the artist supplies what he considers needed, and you can see there is quite a difference "before and after taking." This cockerel was good enough to win four firsts in thirty days at Auburn American Poultry Association Meet, Schenectady, Philadelphia, and Boston, where he won the Championship Cup of the year, still he was so far from perfection that when brought up to the artists ideal, he was unrecognizable, so how can you expect any breeder to sell you birds that will satisfy you after studying the pictures as shown in poultry papers and catalogues.

The Actual Photograph.

The actual photographs of the two New York City Madison Square Garden first prize winners including shape and color club specials, represent as fine specimens as were ever shown and still they are not all we might expect when buying high priced specimens, but they are in fact, better than any breeder that cares to improve his flock would let go at any price, if there was not something more than money at stake in doing so.

The Retouched Photo.

The Standard of Perfection is but the work of common mortals, and after years of unsuccessful attempts to educate the amateur to correct ideals, is still so ambiguous that even the judges interpret its meaning differently, and a large exhibitor has to show all styles and colors of Reds to be sure to have some kinds which are strong in the particular point that the judge may have strongest in his noddle.

It is useless to try and buy a Red guaranteed to win a first prize, as many times (although he might be good enough to be the prize winner) I have seen specimens entirely overlooked by the judge, which were openly pronounced by the breeders present to be the best bird in the show, and I have seen unplaced males shipped direct from the show, C. O. D., for $100 each and prove winners at their next show. So it is more or less a gamble as to where the blue lightning will strike. Still it makes an awful lot of difference to the party who gets it.

While some judges are color mad, still it is generally admitted that shape should be considered above color, and the oblong type of the Reds is one of the strongest characteristic points of the breed, and also one of the most important from the utility standpoint, and should always receive most important consideration when selecting a winner or breeder.

This well developed Rose Comb Cock shows the superiority of the Reds over other breeds, as the traced oblong square would represent the size of body of the average Plymouth Rock or Wyandotte, while the Red shows additional breast and keel, the favorite portion, which all lovers of well cooked poultry desire.

Few specimens of the Red hen family show the emphasized oblong square type of the breed better than this hen, as the oblong block almost stands out to perfection, and her horizontal lines together with lack of curves, gives her the almost ideal body dimension, with the capacity to manufacture eggs for her owner in large numbers all the year round.

A good head for a breeder.

One of the first points considered by the new beginner is the head of any bird. If it has a five point comb it is all right, if not, they consider it no good, thereby proving that a little learning is a dangerous thing. I consider this head about as good as you will find on any prize winner, and also a grand good head for a breeder.

Nearly all pictures are taken from the side, while this front view of the last New York winner shows many very important points that the side view can not determine. Select your birds with wide bodies, well developed breasts, erect carriage of head, and above all, with strong thighs and legs well spread, as in this cut, as this denotes strength and good vitality.

Wyandotte between two good Reds.

I have seen many judges get called down very severely for not giving birds of the type shown in this cut the first prize, while the bird is true Wyandotte type and does not deserve any prize in any kind of competition at all.

While there is some competition between the Single Comb and the Rose Comb varieties, still it is all good natured rivalry, and large breeders who breed both varieties, say there is but little difference outside of the comb, still in some localities one breed is bred almost to the exclusion of the other, so we will let them fight it out. It is a curious fact that the whole World over, the average purchaser when looking for stock or eggs, will pay the highest prices to the breeder who has a strong show record, no matter how this record was secured.

I do not want to belittle the value of prizes honestly won at large or local shows, as they are the real life of the fancy poultry business and without the shows there would not be half the enthusiasm and sport to the average fancier.

What the shows need most of all in the line of reform, is more conscientious judging, by judges free from the influence of the advertising value of the prizes given to certain breeders who patronize their papers. I have seen some horrible examples along these lines.

Every breeder should support his local show and do all in his power to make it a complete success. There is no one person in any locality who does more for the good of the industry than does the Secretary of the show.

I believe all honest breeders do all in their power to give their customers satisfaction, but every poultry raiser is more or less injured by the shortcomings of nature, when stock and eggs are handled by amateurs under adverse conditions.

The greatest scourge of the poultry business in general is the professional solicitors, who at poultry shows and otherwise, by malicious misrepresentations, secure an advertising contract and then print malicious falsehoods, (I have seen breeders lauded to the skies, who did not own a bird fit to breed from) which are an imposition on the reading public. They should have some check put on them.

I hope the amateur will look farther into the case than the mere fact of the ribbons being awarded a certain breeder, as you may discover that the winners have been bought (which is no disgrace to any breeder) or possibly borrowed for the occasion (which is a very common offense) and the breeder has no more birds of their merit to sell you eggs or stock from, and last but not least, the much lauded winners may be expert samples of the fakirs art, and really birds of a quality unfit to breed from. Look up the breeder who has a flock of good birds at home and some to spare for you, when you are buying your foundation stock.

After handling thousands of Reds, I find it most amusing to see how nature never makes a perfect specimen, and that these same disqualified specimens with slightly feathered legs or side sprig combs are the birds that otherwise are nearest perfect, and it is on these worthless specimens that the fakirs get in their fancy work, as when done at the proper time, few judges will throw them out. A disqualified bird has won the American Poultry Association cup in more than one case to my knowledge.

When you attend your local show and feel dissatisfied with the awards, don't be afraid to ask the judge, in a gentlemanly way, to explain what does not seem right to you, and while you may not find out all you would like to know, still it helps to remind the judge that he cannot do just as he pleases without consequences. I have seen some of the old licensed judges act almost as if they really felt they could not make a mistake and that their judgment should not be questioned.

Every judge should disqualify each and every exhibitor who shows birds that have been faked or tampered with, and he should be able to detect these " improvements " or else have it openly understood that all have the equal right to " improve " nature's handiwork, as it is not a " square deal " for the amateur breeder to follow the printed show rules, while the expert exhibitor who knows they are only printed to be broken as he sees fit, goes ahead and remedies the short comings of his birds and wins with his disqualified specimens over the true blue winners which are not properly groomed.

I positively know that for two years the first prize Red cock at one of our leading shows was not shown with tail as nature made it, and another year the first prize males had enough ticking feathers removed from their necks to make even the Barred Rock pickers blush with professional jealousy. Still nothing was done to stop it and the exhibitors winked the other eye, for who cared to go on record as a squealer. It is this kind of work that is killing the enthusiasm of the would-be fancier who, after working hard all year to produce an honest winner, sees his year's work turned down without a fair chance.

There will never be a judge that can satisfy all persons interested and I have often heard good judges criticised by half informed critics who really did not know what made a good Red except possibly size, snut and stamina.

Too many critics take into consideration only the most important points of a bird, while the judge should be a man capable of taking into consideration every line and feather (including those that may have been removed) that goes toward making our ideal, which has never been realized.

I believe words cannot describe the color of Reds. The wording in the Standard is very crude and the small color chart worse than none at all as it is misleading. They say the cocks, the hens, the comb, the eye and the under color should all be red, and you know no two of these are anywhere near the same color, and none of them what is commonly understood to be red.

Popular opinion originally was for anything free from smut to win, later there was a craze for mahogany birds, but today the experience of the largest breeders is that the shade between the two, with as much real redness and brilliancy and little contrast between surface and under color makes the best breeder, and therefore should be encouraged as prize winners instead of giving prizes to birds that are almost worthless as breeders.

The color of hens to resemble the pullets, while very desirable, is seldom if ever seen, and when it does occur is generally accompanied by an argument that there is something wrong, so new beginners must not expect to buy hens looking like the prize pullets you see at shows, no matter what price you pay. The show pullets you see are birds shown at the best time of their lives, generally when about to lay their first egg, and soon after getting down to business will not score near what they did at the show. Red pullets like the pretty girls are "sweet sixteen" but once in their life, although they may be far more practical for the rest of their days.

My color plates give the ideal color nearer than anything that I know of and I will send same to any one sending me 25 cents for DeGraff's Book on Reds.

While judges differ as to their preference in shade of red, still they generally consider the depth of under color throughout all sections, a very important point, and the harmony of the surface color in all sections be it light or dark is a strong winning point, while the health and brilliancy of plumage often decides a close decision in the hottest of classes where many good birds never get a ribbon, for the competition in Red Alley is often getting to be the feature of the whole show, and the wonder of breeders of

other varieties is what makes all Red fanciers such red hot boosters for their favorite breed. If you will but give them a trial you will know why.

To the average breeder the word "Symmetry" means everything or nothing, as they may see fit to take it, and I will show this picture of my famous bird "Amsterdam" as about as good an all round symmetrical male, every point considered, as has been produced, showing a cockerel just as he should be when in perfect feather.

Very often we see an extra large specimen of the breed that, to the average person, looks the perfection of all they could desire, as it surely would fill a very large pot, but from the fancy breeding standpoint, always fight shy of these eleven or twelve pound Red males, as they are not practical and are much better to look at than anything else.

Champion Amsterdam

Often we find cocks that are almost perfect in color in every section, but when fully grown they lack pounds of reaching the weight required, and the accompanying cut shows very plainly this type, which should never be given a first prize, no matter how good the color, and I have never seen any great results, in the long run, come from them as breeders. It seems to be a common thing for the smallest birds to be strongest in color and the largest lacking in color.

Some strains of Reds while having extra good color have run to the game type, as in this cut, having extra long legs, short back, high head, long tail and too much fighting capacity generally. The day has come when we must pay more attention to the type of the birds we are using in our breeding pens.

Then there is the Leghorn type of Reds with long legs, small bodies and extra large combs, and continually crowing, which makes a very poor combination for a good Red, although the color may be all we could ask for. I hope you will see from the above that there is more than color in selecting your choicest birds for the shows.

shown. It is a weak point of the breed, that many males have partly developed tails when on exhibition, due either to lack of growth or being removed to hide the white that came after molting.

While the Standard description for Red Females reads the same for all ages, still it should be explained in some manner that few if any hens ever have the color they had when pullets, and when they do it is usually through lack of strain on their systems from laying. It is an undisputed fact that all hens lose their color after very heavy laying, but who ever heard any complaint that hens layed too many eggs, especially in winter when Reds are right in their element?

I show three ages of the Red hen, with the natural changes of weight as they mature, and the breeding pen shown shows by actual photos my ideal type of what a hen should be for a good breeder.

This bird appears to many to be an extra fine specimen and in nearly every respect he is a grand bird, but he has what is called a pinched tail, having only a few feathers and those all pinched together, which is a very common defect in some strains, and makes a very bad breeder, especially of females.

This cut shows a strong, vigorous cock with his tail carried too low, which is a defect in but few males, as most of them are too high. A good judge should examine very carefully the root of the tail feathers on some birds to see if they have been kinked or broken, so as to give a naturally high tailed bird the appearance of a low tailed one. I have often seen it done.

There is no more pleasing feature of a good male than a perfectly developed tail of the right color carried low, and this bird, while having rather too large a tail, was an extremely fine bird to look at on the lawn.

I believe one of the strongest points, the one which helped "Champion of Amsterdam" win Color Special at Madison Square Garden, was his extremely fine development of the dark glossy green shade on his tail, which few have ever

DeGraff's Book on Reds for 1913 contains many illustrated articles like this, containing facts, which have never appeared in print before, and Poultry, will contain an article each month on same educational lines.

INVEST FIFTY CENTS HERE FOR LUCK

I have made arrangements with Poultry Publishing Company of Peotone, Ill., to act as their Eastern representative, and they will publish an article by me each month, in which I will attempt to give my customers all the interesting and educational matter that I can, to help them make a success with their Reds. This paper is not controlled by any club or association, so I have the privilege of saying whatever I feel is the truth, no matter who it hits, and my sympathies are with the unsophisticated amateurs from now on, as their interests have been sadly neglected. Poultry has sold at $1 per year for over ten years, but I have arranged to take subscribers among my customers at 50 cents each. Send for a sample copy at Peotone, Ill., or better still, send me 50 cents for a copy each month for a year, as there is no other poultry paper published that is as high grade in every respect.

The Saratogian Art Press, Saratoga Springs, N. Y.

PERSONAL VISIT OF L. W. STANDISH, EDITOR OF POULTRY INDEX, STOUGHTON, MASS., AS PUBLISHED IN RHODE ISLAND RED NUMBER, FEB. 16, 1913.

In the beautiful Mohawk Valley, of New York State, in the town of Amsterdam, overlooking the Mohawk river, in sight of the Erie Canal, and West Shore R. R., with the New York Central R. R.'s many trains passing, stands a unique and artistic rubble stone station, for the use of the citizens of the village of "De Graffs".

The fast line of Electrics from Albany to Amsterdam make frequent stops here, and the travelers see from the car windows a beautiful and extensive Poultry Farm that attracts the attention and admiration of the thousands of passers daily. This is the celebrated DeGraff Poultry Farm, known far and wide, as the home of the famous DeGraff Strain of Rhode Island Reds. Here I spent a day recently with the proprietor, Edward T. DeGraff, and the experiences of that day make a story which should be of great interest to the readers of the Index. The Proprietor has the reputation of being one of the best posted breeders of Reds in the country, and I went there to get next, to the man, and his methods, and his birds. Here is a man who has made the Reds a life study. From the time of the original Indian grant, when the country was first settled, this farm has been in the DeGraff family, a family that is known and respected by all the community, and a name that stands for right living, and well doing, carried down from father to son for seven generations on this same land.

Here is no amateur plant. Conditions are all those of an intensely practical, business farm, planned, managed and carried on, on a basis of wide publicity, and in a way that makes it one of the leading farms in the country. The farm is ideally situated. From the railroad tracks rises a gradual slope with splendid Southern exposure, and on which have been built extensive modern poultry houses, and here are to be found some of the very best Reds I have ever seen.

Reds that are distinctly red, true in type, remarkable in coloring, uniform in an unusual degree, and an excellence of shape and standard points, that one seldom finds in any place. It took a long time to go over that hill, to enter the many houses, and see and consider the thousand Reds there on view, and to study the excellent and in many ways unique methods of caring for and breeding these birds, which have a name and reputation all over the land.

His large 600 apple tree orchard, with water piped to all parts, is colonized with movable houses and his methods of caring for same, very simple and effective, in a marked degree. Ample houses, splendid yardage, modern methods of feeding, open fronts, trap nesting and the segraruon of pens, which are all the concomitants necessary to present day success, in raising exhibition and fancy stock, and the visitor be he a neophite or poultry sharp, can consider the ways and learn much that is valuable and worth while.

It is seldom that one can see one thousand birds all of one breed and color and then come away, as I did with a distinct impression of general excellence that drove from my mind the impression of individual excellence and substituted therefor the overwhelming sense that all birds on the place were good. Had I my wish, to go back to this farm and carry away 50 or 100 birds, the finest of the flock, I confess that I would be bewildered as to where to begin and where to stop in the task of selection. After one has for half a day wandered through pen after pen of rare, noble birds, and has brought to his notice scores of what he would declare were equal to prize winners in any show, one gets a sensation akin to that which befalls him after days of wandering through art galleries, and being surfeited with noble paintings and works of art. Edward T. DeGraff—Here is a man who not only knows Reds, but who can talk Reds, who can write about Reds, and who is competent and is now engaged in task of reviewing for one of the leading poultry papers of the country, the work of the best judges in the leading poultry shows of the land, and he is doing it too, so well as to attract the attention and interest of poultrymen by the keenness of his observations and by the force of his criticisms, and the value of his conclusions.

A long time ago Mr. DeGraff got in bad with the powers that then were in the A. P. A. It is a long story and we do not need to go into it, only to say that a careful investigation of the matter shows conclusively to us, that there was a mixture of A. P. A. politics, some jealousy, a little bumptiousness of a man he had assisted in many ways, and a lot of desire to make some one the "Goat" for the protest of the times against the then almost universal practice, carried on at the shows. Mr. DeGraff became the "Reform Goat". For a time the sins of the many were visited on his head. Unlike most men, the victim was not easily downed, or daunted. He went his way, told facts as they were biding his time, and lo and behold here stands a man, who in spite of the many stones which have been thrown at him, still has continued to become greater and greater, as a breeder, whose business has grown by leaps and bounds, and who today is considered to be one of the best abused men in poultrydom, and for whom the reaction has come, and will do him far more benefit, than any injury that has resulted from this unprecedented action in poultry circles.

Why? Because he has proved that he is honest, and straightforward with his patrons, because he is raising, and selling, birds that few can equal, and giving dollar for dollar to customers of years acquaintance, and who is receiving the unique testimonial, from the very men who have condemned him, that perhaps they were mistaken and too hasty in their judgments. Let it rest. Mr. DeGraff himself is his own best answer to the charges.

There is no doubt that he is doing today a very extensive business. On his desk while I was there I saw at least 25 orders for immediate shipment, one for a shipment of 200 eggs to one man and another for a trio of birds at $100. Few breeders can give their customers what DeGraff can.

Mr. DeGraff always was a Red enthusiast. He also believes in printers ink, and that too of the best kind. His catalogue is a de luxe affair for which thousands of breeders are glad to pay 25c.

It is made up in a beautiful way with magnificent colored plates fit for framing in the office of any poultryman, it gives much valuable information in regard to poultry culture, and is well worth the price he asks for it. The new edition, I was allowed to inspect in the proof, and it is to be a great work of art.

Taking into consideration the present facilities in houses and yards and the many additions now in process of construction, together with his invincible stock of over 1000 of the finest of Reds together with his established reputation all over the World of breeding the best Reds in America, the whole forms a combination that it will take other breeders years to equal.

He has facilities for doing an unlimtied business and he is never satisfied.

Mr. DeGraff's monthly articles in "Poultry" are fast becoming appreciated as the best unprejudiced "up-to-the-minute" writing on matters pertaining to the Reds.

Besides his own farm Mr. DeGraff has other red farms under his control where reds are raised for him from his stock, so that he can supply promptly all demands that may be made on him for hatching eggs or stock.

I visited one farm where I saw 800 healthy utility flock bred from the very finest of stock, where they were averaging 350 eggs a day in coldest weather of the season, proving that utility and exhibition qualities can be combined in one strain.

I like DeGraff. Enemies he has. But I will say this about him. He knows the Reds. He has the best Reds. He satisfies his customers. He is energetic, straight-forward, frank in regard to his past, is not bitter and he has come out on top. After all the man who wins out against adverse circumstances has more call for my admiration than the one who never has perhaps erred, because he has never been tried in the balance. If you want Reds you should consider DeGraff's Reds and by all means send for his catalogue. It is a beauty and a joy forever.

L. W. STANDISH,
Editor Poultry Index, Stoughton, Mass.